NASA SP-2000-4523

The Eclipse Project

by
Tom Tucker

NASA History Division
Office of Policy and Plans
NASA Headquarters
Washington, DC 20546

Monographs in
Aerospace History
Number 23
2000

Library of Congress Cataloging-in-Publication Data

Tucker, Tom, 1944-
   The Eclipse Project/by Tom Tucker.
     p. cm. — (Monographs in aerospace history; no 24) (NASA history series) (NASA
   SP-2000 ; 4524)
   Includes bibliographical references and index.
   1. Eclipse Project (U.S.)—History. 2. Rockets
(Aeronautics)—Launching—Research—United States. 3. Towing—Research—United
States. I. Title. II. Series: III. NASA history series IV. NASA SP ; 4524

TL789.8.U6.E42 2000
629.43'2'072073—dc21

                                            00-067616

For sale by the Superintendant of Documents, U.S. Government Printing Office
Internet: bookstore.gpo.gov Phone: (202) 512-1800   Fax: (202) 512-2250
Mail Stop SSOP, Washington, DC 20402-0001

Table of Contents

Foreword

The Eclipse Project by Tom Tucker provides a readable narrative and a number of documents that record an important flight research effort at NASA's Dryden Flight Research Center. Carried out by Kelly Space & Technology, Inc. in partnership with the Air Force and Dryden at Edwards Air Force Base in the Mojave Desert of California, this project tested and gathered data about a potential newer and less expensive way to launch satellites into space. Whether the new technology comes into actual use will depend on funding, market forces, and other factors at least partly beyond the control of the participants in the project. This is a familiar situation in the history of flight research.

Frequently, the results of discoveries through flight research are not implemented immediately after projects are completed. A perfect example of this phenomenon is the lifting-body research done in the 1960s and 1970s that finally lead to new aerodynamic shapes in the world of aviation and space only in the 1990s. Even then, the lifting-body shapes (for the X-33 technology demonstrator and the X-38 prototype crew return vehicle) were only experimental. Other technologies emerging from flight research, such as movable horizontal stabilizers, supercritical wings, winglets, and digital fly-by-wire moved more rapidly into actual use in operational flight vehicles, but it was never crystal clear at the start of a flight research project whether the results would simply inform future practice or would be adopted more or less completely by air- and spacecraft designers.

Regardless of the eventual outcome in the case of the Eclipse Project, it was a unique and interesting experiment that deserves to be recorded. Tom Tucker has told the story in an interesting way that should make the monograph a joy to read. I thank him for his hard work, writing skill, and his flexibility as the monograph went through the coordination process. He was busy with teaching and writing another book; yet he unfailingly responded to my requests for technical changes in the monograph as various participants refined the details of the events surrounding the tow testing of the QF-106 behind its C-141A tow vehicle.

As editor of the monograph, I also want to express my appreciation to Jay Levine for his expert work as layout artist and to Carol Reukauf, Mark Stucky, Al Bowers, Bob Keltner, Fred Johnsen, and Bill Lokos for their comments on the drafts of the study. Their assistance has made the account much fuller and more accurate than it could have been without their taking time in very busy schedules to apply their personal knowledge and expertise to the text at hand. I recommend the result to anyone interested in the history of aviation and space technology. It will be especially valuable to anyone undertaking tow testing in the future.

J. D. Hunley, Historian
NASA Dryden Flight Research Center
5 December 2000

Star

When I was writing this history of a space technology issue, I was also busy at work on a book about Benjamin Franklin's lightning science and deeply immersed in 18th century culture. In pursing the 18th century project, I found myself seeking out dust-covered documents from archives. What a relief it was to research the Eclipse story whose participants were all among the living and so willing to provide information. Eclipse was a joint effort uniting the efforts of three agencies, and to them all I owe a debt of gratitude, to Kelly Space & Technology, the U.S. Air Force, and NASA Dryden Flight Research Center.

Another Enlightenment perspective guided my approach to the Eclipse project. In the 18th century, scientists and technologists focused on certain key issues, such as a practical means of finding longitude at sea and locating the Northwest passage. The solutions to these problems would have dynastic import for nations, and individuals or groups finding practical solutions would earn a fortune worth a king's ransom.

In our day, there has been a similar key issue, one just as important to the course of human history, just as potentially rewarding for those who find the solution. In the aerspace industry, it has become a sort of invisible barrier. For more than twenty years, the cost for space launch has remained about $10,000 per pound. No innovation has appeared to solve this problem. Among the scores of creative, exciting ideas conceived by small start-up companies trying to meet the challenge was the Eclipse project. Behind the tiny Eclipse project resonated a large issue.

I owe a great debt to the many individuals, programs, and organizations which enabled me to write this history. First, I am grateful to the NASA-ASEE Summer Faculty Fellowship Program which brought me to NASA Dryden Flight Research Center out in the Mojave Desert and supplied me with every kind of support needed for research and writing. At Dryden Center, Don Black and Kristie Carlson provided much courtesy and good advice. At the Stanford University Department of Aeronautics and Astronautics, Melinda Francis Gratteau, Program Administrator, and Michael Tauber, Co-director of the Program, aided me invaluably with their help, consideration, and provision of opportunities. The programs which brought me in contact with scientists and engineers who were NASA ASEE fellows at the NASA Ames Research Center helped me in thinking about and clarifying this invention-history project.

Many people inside and outside the three participating agencies gave generously of their time and expertise in interviews and correspondence. These included: Bill Albrecht, Mike Allen, Don Anctil, Bob Baron, Al Bowers, Tony Branco, Dana Brink, Robert Brown, Randy Button, Bill Clark, Mark Collard, Bill Dana, Dwain Deets, Casey Donohue, Bill Drachslin, Ken Drucker, Roy Dymott, Stuart Farmer, Gordon Fullerton, Mike Gallo, Joe Gera, Tony Ginn, Ken Hampsten, Stephen Ishmael, Mike Kelly, Bob Keltner, Kelly Latimer, Bill Lokos, Mark Lord, Trindel Maine (again), Jim Murray, Todd Peters, Bob Plested, Debra Randall, Dale Reed, Carol Reukauf, Wes Robinson, Kelly Snapp, Phil Starbuck, Mark Stucky, Gary Trippensee, Daryl Townsend, Mark Watson, Roy Williams, and Joe Wilson.

Readers of drafts along the way offered many valuable comments. I especially thank: Al Bowers, Fred Johnsen, Mike Kelly, Carol Reukauf, and Mark Stucky. I am grateful to Dennis Ragsdale and Erin Gerena of the NASA Dryden Library for tracking down my numerous research requests. Steve Lighthill, Jay Levine, and the NASA Dryden Graphics Office as well as the NASA Dryden Photo Lab went above and beyond the call of duty in giving this project the benefit of their talents. Steve Lighthill also deserves recognition for his expert work arranging for the printing of the monograph through the Government Printing Office.

Last and most, I owe a debt to Dill Hunley, editor, historian, advisor, facilitator, friend, and when a phrase needed a different turn in the face of an impossible deadline, co-author. He made this book much better than it started.

Tom Tucker
Rutherfordton, NC
12 December 2000

Eclipse

The Eclipse Project

Tom Tucker

Start Up

If you wander the halls and look in office spaces at NASA's Dryden Flight Research Center in the Mojave Desert northeast of Los Angeles, you'll see that nearly every researcher has walls decorated with mementos from projects completed. Trophies, keepsakes, awards–these often take shape as photographs. Hundreds of projects have resolved on these walls into 8-by-10-inch glossies.

When you look in some offices, however, you see what looks like a yachting trophy. It's a snippet of heavy-duty rope. It is installed on a generic memento plaque, but it was also recently the centerpiece in a futuristic project brought to Dryden by a small venture company named Kelly Space & Technology, Inc. This company hoped to demonstrate a new approach to satellite launching by first towing a space launch vehicle to altitude behind a transport airplane.

Aerospace engineer Jim Murray keeps a unique memento of his participation in the aerotow project–a large, messy jumble of monster rope that dangles from the ceiling. It's the only trophy in his office space. When Murray leaned forward one day under the fluorescent glare, rubbing his hand back through an unruly mop of hair so that he reminded me of the inventor in the movie *Back to the Future*, he preferred to talk of his current assignment, designing an airplane to fly the atmosphere on Mars. But I couldn't help noticing the rope over his shoulder.

It is an eerie, snarled trophy–utterly unlike the polite snippets of rope that decorate other offices. In a glance, you can see the lengthy strand in an entanglement no human fingers have devised. You assume correctly that it represents the aftermath of some violence thousands of feet overhead in the desert sky–a rope that has outslithered any mathematical prediction, a mesh of energies, a witness to unknown forces.[1]

If you stare too long, you imagine the scent of jet fumes, the deafening roar of engines, the rope itself powering off the wall toward you. It represents a curious project, one that generated controversy at the research center and at NASA Headquarters in Washington, D.C. This was so because its central technology, for all the leading-edge electronics and aeronautics developed around it, was the rope. NASA involvement began in 1996 when a small start-up company first approached NASA Dryden from just over the next mountain range in San Bernardino.

Mike Kelly, founder of Kelly Space & Technology (KST), is a pleasant-looking man in his mid-forties with graying hair, and when he speaks, he often brings both hands up as if trying to frame an idea in midair. He has an engineer's hesitation when he starts talking, which soon disappears as his enthusiasm takes over. He remembers when he had the tow launch idea. It came to him late in the winter of 1993. He was working out of his home office, then in Redlands, California, just after he and TRW had parted ways. For some time, he had been thinking about a problem in the communications industry: the stiff costs of placing satellites into orbit. Despite the rapid growth of Internet and telecommunications technology, despite many breakthroughs in efficiencies that had lowered costs, there had been no breakthroughs in the satellite delivery system. The high cost of launch had not changed for several decades. It is difficult mathematics to estimate exact costs for this service with its federal subsidies, but a launch price tag might come in near $10,000 a pound.

"I was sitting at my desk," recalls Kelly. "I had been thinking for a long time

---

[1] James Murray, interview by author, 14 June 1999, and the author's observations during it.

about strategies for taking off from the ground with a reusable rocket." During one period at TRW, he had investigated reusable launch vehicles, RLVs the industry calls them. He thought about the Shuttle approach, how the piggyback worked, and he thought of Pegasus, how the under-wing stowing worked. And then he thought of pulling gliders on a rope.[2]

The moment was the genesis of his project. Curiously, he recalls no excitement at the moment, merely a sense of one hazy concept among many possibilities to file away for later evaluation. "But I went for a walk," he says, "and the towing idea came back, and I began saying to myself, 'you know this makes a lot of sense,' and the ideas began to come fast and furious. By the time I got back to my desk, it had me."

If you keep adding weight to a space launch vehicle, reasoned Kelly, to get more thrust you add more propellant–which adds more weight and adds greater operating costs. But Kelly's concept–and it was a leap for an engineer/manager who had devoted his career to ballistic missiles–was to adapt to space launch technology what was essentially the technology of a glider towed on a rope.

It takes formidable engine thrust to get a launch vehicle to 20,000 feet. Kelly reasoned, why not let a transport airplane do all that first-stage work? Kelly next pursued a bit of research in the San Bernardino Public Library and discovered historical precedent. He found that in the 1920s a British woman, the romance novelist Barbara Cartland, had addressed the same problem because she wanted fresh vegetables from the Continent on her plate. At the time, airplane people explained to Cartland they did not have a technology for carrying vegetable cargoes. Although their airplanes could carry the weight, the problem was low density. The airplanes could not carry the volumes of something like French lettuce, for instance, that would make the cargo profitable. She had replied, why not pull a big airplane with plenty of volume behind a small airplane with an engine? A glider on a rope offers a simple way to transport more volume (and more weight).

"You can pull more than you can carry," says Kelly. The point can be intuitively grasped without understanding airplane lift and thrust. Consider, for example, moving heavy boxes. Consider carrying the load in your arms and walking. Consider instead putting the boxes on a sled and pulling the sled by rope on snow. The difference is rocket launch versus tow launch.

As Kelly's idea grew, its efficiencies seemed to multiply. For example, where a space launch pad might cost as much as $75 million to construct and is expensive to maintain, Kelly's idea depended on a conventional airport runway. Where one-shot rockets are costly disposables, Kelly envisioned his transport and his second-stage vehicle returning home to the airport. Where weather conditions imposed costly delays on launch pad takeoffs, Kelly's approach offered flexibility in departure site and scheduling.

The ideas flooded around him on that brisk late-winter afternoon. In terms of space launch, he had moved from the ballistic missile paradigm to the commercial airline paradigm. By the time he approached the sidewalk to his Redlands home, Kelly had covered quite a bit of ground.[3]

---

[2] Mike Kelly, interview by author, 16 July 1999. The Shuttle launches piggyback, so to speak, on its external tank with two solid-rocket boosters attached. Pegasus launches from under the wing of an L-1011 (initially a B-52) launch aircraft.

[3] Kelly interview.

* * *

A few years later, Mike Kelly turned up at NASA Dryden with his experiment. In the interim, he had formed his company, Kelly Space & Technology, found partners and investors, and hired a small team of engineers, many of them retirees from aerospace enterprises in the San Bernardino valley. He had filed a patent application for his winter afternoon brainstorm. "Space Launch Vehicles Configured as Gliders and Towed to Launch Altitude by Conventional Aircraft" he called it, and the patent was later granted on 6 May 1997.[4]

After six months in business, KST had encountered a kindred spirit on the issue of low-cost access to space. He was Ken Hampsten of the Air Force Phillips Laboratory,[5] who had published a new topic for SBIR, Small Business Innovation Research, a broad federal program that encourages groundbreaking and creativity in small companies. That year Hampsten asked for proposals to be submitted in the area of space launch technology. In April 1995, he chose KST from more than thirty applicants and gave it funds for a Phase I SBIR grant, a feasibility study on paper. With that success behind them, the Kelly people next applied for and received a Phase II SBIR grant for a study that would be a demonstration of concept in real flight. Kelly wanted to do a subscale demonstration of bigger things that lay ahead. He wanted to take off and tow a high-performance delta-wing aircraft behind a transport aircraft. His hope was an alliance. The Air Force Flight Test Center (AFFTC) at Edwards Air Force Base (AFB) would supply and fly the transport (a C-141A). The towed airplane would be lent or bailed from

another Air Force unit, and NASA Dryden would contribute its flight research expertise.

From the start, Eclipse flight issues divided experts at Dryden. Would the rope introduce some new and possibly dangerous dynamic to the airplanes? The KST visionaries and many of the Dryden people, who were recreational glider pilots and had experience being towed on a rope all the time, saw no problem. One of the early project managers, Bob Baron, addressed this issue in the cover designs on Eclipse reports. He had an artist introduce images of the transport in front of the interceptor and then draw a white line from the tail of the C-141A to the nose of the F-106 to represent the rope. Ultimately, the rope path proved fascinatingly different. But at the time, there was no available evidence to the contrary. Baron reduced the problem for his report-readers. He reduced it to a reassuring straight line.

There arose a growing suspicion, however, among many engineers and pilots at the center, within and outside of the project, that the hazards were not as minimal as those attending recreational gliding, not so negligible as to be reduced to a straight line–that somehow dangling a 30,000-pound Cold War interceptor on a barge rope might be dangerous.

Curiously, there was little literature on the subject. There existed no validated modelings of towed flight reality. Research through the library at Dryden initially turned up the pioneer Anthony Fokker patenting tow technology in 1919, misty accounts of extensive German aerotow experimenting before and during World War II, and some brief accounts of the United States working on the WACO

---

[4] Mike Kelly, United States Patent 5,626,310, "Space Launch Vehicles Configured as Gliders and Towed to Launch Altitude by Conventional Aircraft," 6 May 1997 (See document 2 of this monograph).

[5] Redesignated the Propulsion Directorate of the Air Force Research Laboratory in October 1997.

3

glider.[6] Mostly, the research turned up anecdote.

The anecdotes did not bode well.

One account came from the legendary Royal Navy Test Pilot, Captain Eric Brown. Rogers Smith, who was then Chief Pilot at Dryden, had a personal connection to Brown and asked for his input. Brown had flown the German-built Me 163A and Me 163B when they were towed in flight tests by Spitfires. He wrote, "If the tug's slipstream was inadvertently entered, a very rough ride ensued and control was virtually lost until the towed aircraft was tossed out of the maelstrom."[7]

During the same period, a B-29 towed the Me 163 at Muroc Army Air Field (now known as Edwards AFB). Brigadier General Gustave Lundquist–writing later about the experience–stated, "This sounds simple enough, although it was anything but. In fact, it was the scariest experience I have ever encountered in all my flying."[8]

The desert base had more history to offer. Several older Dryden pilots had flown wake turbulence tests in the 1970s and witnessed Cessnas and Learjets tossed upside down as if they were toothpicks by the wake of Boeing 747s. There was the case of test pilot Jerauld Gentry who flew on tow in the lifting-body program and twice rolled over on tow release.[9] Perhaps the earliest local anecdote concerned a tow crash in September 1944. The test pilot had walked away unscathed, and the incident–reported in a sort of deadpan, gosh-gee-whiz, 1950s style by eyewitnesses in their sworn statements–assumed the proportion of comic legend on the base.[10] But the story of a nylon rope rubberbanding back at the towed airplane seriously concerned the Eclipse investigators.

They felt even more troubled by the accounts from Europe. There was the incident involving the Germans who suffered 129 deaths in a 1941 towing accident. The ropes to their vast glider, the Gigant, snarled in a crash that made aviation history.[11]

The Elements

Kelly planned to use a modified Boeing 747 for his ultimate tow plane. No expensive design, no lengthy develop-ment, no vast web of flight qualification testing awaited KST. The towed airplane was named the Eclipse Astroliner, and it

---

[6] James E. Murray, Albion H. Bowers, William A. Lokos, Todd L. Peters, and Joseph Gera, *An Overview of an Experimental Demonstration Aerotow Program* (Edwards, CA: NASA TM-1998-206566, 1998).

[7] Eric Brown, personal letter, 17 June 1997.

[8] Gustave E. Lundquist, "From the PT-3 to the X-1: A Test-Pilot's Story," ed. Ken Chilstrom and Penn Perry, *Test Flying at Old Wright Field* (Omaha, NE, 1993).

[9] R. Dale Reed with Darlene Lister, *Wingless Flight: The Lifting Body Story* (Washington, DC: NASA SP-4220, 1997), pp. 60-62. The phrase "on tow" simply means that the aircraft was being towed by another vehicle.

[10] U.S. War Department Report of Aircraft Accident, no number, (Moffet Field, CA, 5 September 1944).

[11] The Gigant's technical designation was the Me 321. It was a large glider aircraft that could be towed by a single large aircraft or up to three twin-engine aircraft. The Discovery Channel has shown a video of the glider accident on its *Wings of the Luftwaffe* series produced by Henninger Video, Inc. See also *Jane's 100 Significant Aircraft, 1909-1969*, ed. John W. R. Taylor (London: McGraw-Hill, 1969), p. 108, and especially William Green, *The Warplanes of the Third Reich* (Garden City, New York: Doubleday and Company, Inc., 1970), pp. 645-648. Thanks to Al Bowers and Fred Johnsen for guidance to the sources listed here.

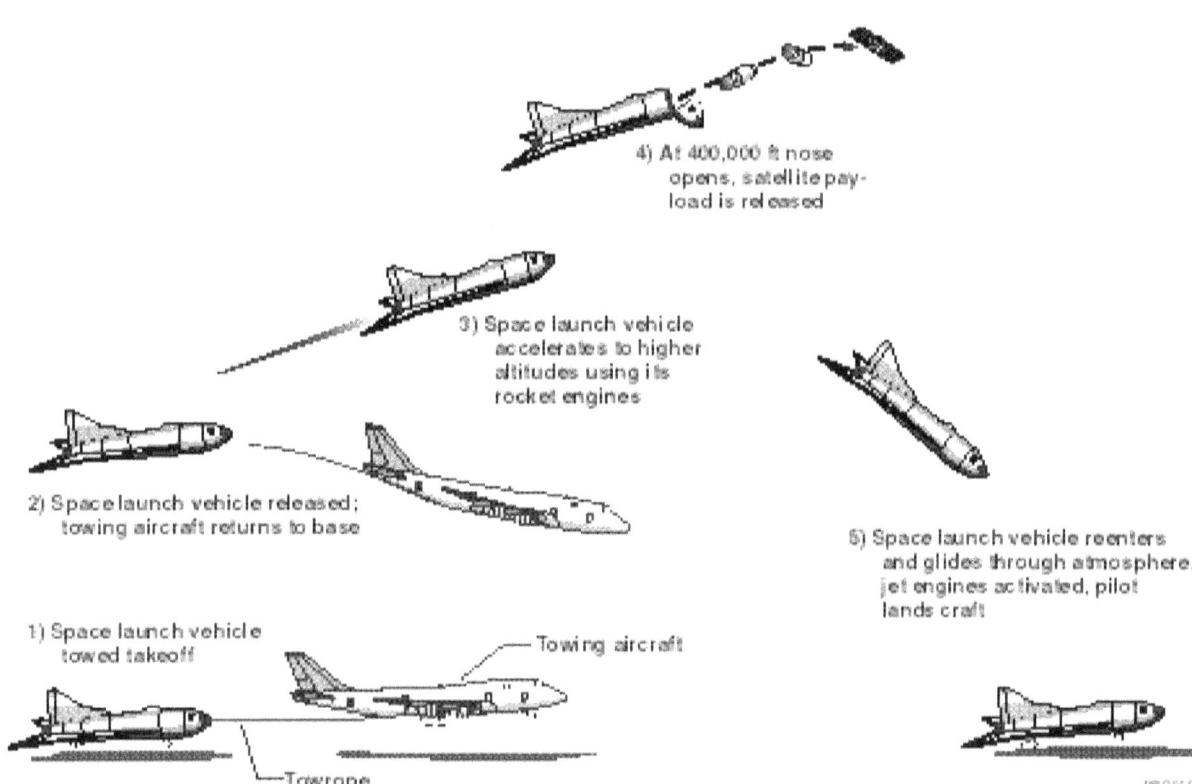

4) At 400,000 ft nose opens, satellite payload is released

3) Space launch vehicle accelerates to higher altitudes using its rocket engines

2) Space launch vehicle released; towing aircraft returns to base

1) Space launch vehicle towed takeoff

Towing aircraft

Towrope

5) Space launch vehicle reenters and glides through atmosphere, jet engines activated, pilot lands craft

Figure 1. Aerotow space launch concept schematic. (Design 980440 by the Dryden Graphics Office)

was a conceptual re-combination, with its essential element the wing root and fuselage of another airplane, the Lockheed L-1011, and its motors, flight-control systems, instrumentation, and thermal protection all borrowed from other current and flight-qualified aircraft.

To test this concept, Kelly needed airplanes, vehicles whose identities were at that point unknown. Because the Astroliner was a delta-winged vehicle, Kelly sought out a delta-winged intercep-tor so that it would provide a proof of his concept using an aerodynamically similar towed vehicle. He also was looking for a transport aircraft that would be a scaled-down version of the airplane that would tow the Astroliner in his concept.[12]

Kelly needed a transport airplane, an interceptor airplane, pilots, crews, flight-test engineers, and a rope.

* * *

From the beginning, the rope was there: Patent 5,626,310, column 7, paragraph 2 of Kelly's claim text, "The launch vehicle would be coupled to the tow aircraft by a flexible cable. . . ." The rope was destined to become part of space technology, but its pedigree dated back over the centuries. The apprehension about the concept was there in the beginning, too. Column 7, paragraph 2 of the patent continued: ". . . the cable . . . would be attached to the aircraft . . . at or near the tow aircraft's center of gravity. This is done to minimize the overturning moments which would be applied to the aircraft by the tow line."[13] And as he and his partners discovered, aeronautical data on overturning moments generated by tow-rope configurations turned out to be nil.

The primary objective of the Phase 1 SBIR study had been to define a basic

---

[12] Comments of Mark P. Stucky, project pilot, on the original draft of this study, 16 September 1999.

[13] Kelly, patent.

system for a tow demonstration including the tow and towed aircraft, tow rope, test criteria, and operational procedures. One of the more critical tasks was the selection of a towline. Planners investigated four materials: high-strength steel and three synthetic-fiber ropes–Kevlar®, Spectra®, and Vectran®. Tracor Aerospace, a Phase 1 sub-contractor, recommended Vectran® as a result of the company's experience in towing targets.

Rope is old technology, dating back to ancient Egypt. Rope of earlier centuries was hemp, and the earliest ropes were hand-woven with strands no longer than the six-foot lengths supplied by bushes along the Nile. When KST Manufacturing Manager Roy Hofschneider went looking for a Vectran® vendor, he discovered a small New York-state supplier, Cortland Cable, which had primarily produced high-test fishing line but then branched out into the manufacturing of rope for barge towing. Ultimately, Cortland Cable would supply the project with 1,000-foot lengths of a synthetic rope, every strand woven in and never broken or spliced but continuous from end to end, as specified by the Eclipse team.

Vectran® was, indeed, an amazing material. It was a liquid-crystal polymer fiber with many virtues. It had the qualities required for the difficult task at hand, including strength, the ability to damp vibration, minimal inclination to absorb moisture, high dielectric and chemical resistance, a high melting point, strong disinclination to degrade in extreme temperatures, and great ability to withstand the effects of abrasion. The other synthetics shared many of these attractions, but Vectran® offered the best match with operational requirements.

From the standpoint of cost, steel was a tempting choice, but a steel cable of equivalent strength would weigh five times as much as Vectran®. The large strength-to-weight ratio and resistance to temperature degradation decided the Eclipse team on Vectran®.[14]

As a shock absorber of dangerous oscillations, nylon had appeal. Not only was nylon of interest because it could damp energy exchange between aircraft, but the Air Force already had extensive experience with nylon rope (when a C-141A at Edwards set the world record, 70,195 pounds for heavy cargo drop, it extracted and dropped the load on nylon chute lines).[15] But nylon was good and bad–it damped energy, which was good, *and* stored energy, which was bad. And unfortunately, nylon weakened as it was stretched. In effect, it destroyed itself, the fibers actually cutting one another.

The Vectran® rope, on the other hand, got stronger when stretched–at least the first time. In fact, an initial stretching of the rope became part of every Dryden flight preparation. Vectran® had interesting abrasion qualities, too. When the polymer rope began to wear, it fuzzed up on the outside and thus protected the inner rope from wearing. Yet despite the rope's great strength, Vectran® also had a weakness–it was vulnerable to sunlight. After the ropes were prepared for flight, the crew had to find a closed storage area where it could safely store the puzzling rope, which was used only for one flight per 1,000-foot length.

* * *

From the start, Kelly's concept required a big tow airplane. It had to be a real brute. In his patent under "Summary of the

[14] Above three paragraphs based upon comments provided by KST on coordination, 15 November 2000.

[15] Mark Watson, interview by author, 29 June 1999. Robert Brown of Lockheed Martin confirmed that a C-141A had dropped a sequence of loads weighing a total of 70,195 pounds at El Centro Naval Air Station in July 1965 by calling the Air Mobility Command History Office, whose archives contained that information.

Side view of the C-141A tow aircraft. (NASA photo EC98 44391-25 by Carla Thomas)

Invention," Kelly explains, "The tow aircraft contributes only thrust, not lift, to the launch vehicle."[16]   The tow plane had to have power and deliver it during the critical milliseconds of takeoff.

The CV-990 first gleamed with promise as a tow aircraft. The Kelly engineers were intrigued. Although they knew the transport had some performance shortcomings, there was a CV-990 at NASA Dryden Flight Research Center already instrumented for research but at the time devoted to testing Shuttle tires. KST negotiated to use this aircraft but could not gain access. Where could it find a testbed?

A C-141A Starlifter rested on the ramp at Edwards Air Force Base. This airplane and its ilk had been workhorses for the Air Force for a generation. They were not fancy transports. The C-141A crew knew this particular vehicle very well. It possessed a special history and had even set a world's record for heavy cargo chute drop. The airplane bore serial number 61-2775 and was the first to roll off the assembly line. It was a pre-production model devoted to testing. Although the airplane had logged a mere 10,000 hours, its days were numbered. A calendar date would soon arrive requiring perhaps more than a million dollars in maintenance expenditures, which would not be forthcoming. The transport with the illustrious history was itself about to become history.

Capt. Stuart Farmer, the Air Force C-141A test pilot on the Eclipse project, compared the transport he flew to the B-52 in the sluggishness of its response. No finesse was there–or ever intended. "As far as roll and pitch control [were concerned]," he grinned, "it's kinda deadbeat."[17]   But the airplane had power. In the equation of operations which Kelly had sketched, in the part of the equation that represented thrust, this was, as Air Force Loadmaster Ken Drucker later explained, "one overkill airplane."[18]

---

[16] Kelly, Patent 5,626,310, column 4, paragraph 4.

[17] Stuart Farmer, interview by author, 25 July 1999.

[18] Ken Drucker, interview by author, July 1999.

Kelly negotiated with the Air Force for months. At one point, he received an offer of "limited support." Of course, when someone comes along asking for a four-engine jet transport, flight crew, maintenance crew, airplane modifications, and instrumentation, to offer "limited support" is one way of saying no. The next months resulted in intense negotiation and leveraging.

Curiously, the skepticism about Eclipse may have kept the project afloat. Because the project was viewed in various Air Force units as so underfunded, so unlikely, no one took the responsibility for killing it off. Eclipse continued to survive.

At some point in the summer of 1995, Eclipse established a relationship with Dryden. There were meetings with Gary Trippensee, who would be assigned as the first NASA project manager, and Stephen Ishmael, who was advising the project from a pilot's point of view. Nowhere did Dryden pledge flight safety responsibility. Nor did Dryden offer a pilot to fly. Eventually, this situation would change. However, at the time, Ishmael received an assignment to a management position with the X-33 project involving a prototype for a possible future launch vehicle,[19] and Eclipse was given a different pilot.

* * *

Pilot: in column 4, paragraph 2 of the Patent under "Summary of the Invention", Kelly described the towed plane as having "a control system which permits it [to] fly either autonomously or under remote control."[20] In the final version of the invention, the pilot would be optional. But for test flight, Kelly needed a real research pilot very badly.

"I was the new kid on the block," says Mark Stucky, a young former Marine test pilot who came to the Dryden research pilot's office early the next spring. He had the trim build all the pilots do, green eyes, and an expression somewhere between politeness and amusement. He arrived with a nickname, Forger, that had nothing to do with aeronautics, which in fact dated back to some obscure event in his college days, but instantly, it seemed, the whole base knew him as Forger.

Coincidentally, more than a year before, Forger had a glimpse of the Eclipse proposal. It was at NASA's Johnson Space Center (JSC) in Houston, Texas. His boss had called the former Marine into the offices to look at some papers from Kelly. His supervisor knew he had years of experience flying gliders and sailplanes on tow and wanted his thinking on the feasibility of aerotow involving jets. At the time, the possibility that Forger might ever work for NASA Dryden, let alone pilot Eclipse, seemed as remote as flying around the rings of Saturn. He thumbed the neatly drawn pages. What did he think, inquired his boss as they stared at a drawing of the pilot in the airplane pulled by a rope.

"I would love to be that guy," Forger thought.

If you ask him now, Forger tells you he was assigned by Dryden as project pilot on Eclipse "because no one thought it would happen."[21] In February 1996, his assignment to Dryden offered him a

---

[19] In a conversation with NASA Dryden Chief Historian J. D. Hunley, Ishmael indicated that during this period of negotiation, he contemplated the possibility of becoming the project pilot as an employee of KST. As suggested in the narrative, this never came to pass.

[20] Kelly, Patent 5,626,310.

[21] Interview of Mark Stucky by author, 15 June 1999.

chance to accumulate some local-style project experience, if only in the meetings and briefings.

Joe Wilson remembers watching Forger fly the F-18 High Angle-of-Attack Research Vehicle. Wilson, a controls and handling qualities engineer, is a sandy-haired, tall man with eyes that gleam with curiosity, who functions at Dryden as the Boswell of the center.[22] Over the years, he has kept journals, partly on computer, in which he records the daily events at Dryden, nothing by way of official report, but personal notes on what he has seen and heard in this almost small-town community of experimenters.

Wilson remembers watching Forger flying spin tests, acrobatic descents from 40,000 feet and then afterwards tracking tests where he followed another airplane at high speed and through abrupt rolls, trying to keep the airplane in his gunsights. No matter what the other pilot did, he was in Forger's crosshairs. "When you see a smooth trace on that," says Wilson, "you know you've got a good pilot." How good was Forger? Wilson's eyes get big.

"*Very, very* good," he nods his head.

But there's a tricky paradox confronting research engineers, Wilson says. "Smooth pilots can lead you down the primrose path." He explains that there are two piloting styles. "There are low-gain pilots," he says, "and high-gain pilots. A low-gain pilot–if you look at the charts–seems barely to touch the stick, almost as if the airplane is flying itself. A high-gain pilot is working the stick constantly,

giving it inputs the whole time."[23] But that day as he scanned control strips, he realized Forger, upon request, could be either.[24]

Yet as Forger established a reputation at the center that spring, Eclipse flight remained unlikely. A plane had not yet been identified. The KST engineers knew that many airplanes might serve as the towed vehicle. They preferred a delta wing. That is, they preferred the wing of a Space Shuttle, the shape that enables reasonable handling characteristics when the airplane descends from space into the atmosphere. Over at KST, one of the company's major investigators, engineer Don Anctil, came up with the idea that they might be able to use an airplane that was nearing the end of its operational days decaying in the humid, sweltering Florida subtropics. This was the F-106, which Anctil had worked on years ago as a young structural engineer at Convair in San Diego.

The F-106 was a remarkable airplane. It had an incredibly robust structure, beautiful clean lines, and power to spare. If you asked the Air Force pilots who flew and serviced the old warrior, they smiled–it was a Cadillac; they loved it; they had a soft spot in their hearts for it. They bestowed upon it the affectionate nickname, "Six."

The F-106 was born in the mid-1950s, an all-weather interceptor created to defend the country from enemy weapons systems. It still holds the official world speed record for single-engine aircraft, 1,525.95 miles per hour set at Edwards AFB in 1959.[25] Pilots remembered it as a

---

[22] James Boswell was the biographer of Samuel Johnson. His name has become a synonym for an admiring biographer or chronicler.

[23] Joe Wilson, interview by author, 22 July 1999.

[24] Joe Wilson, interview by author, 28 June 1999.

[25] According to KST reviewers of a draft of this monograph. Of course, this has to be qualified to air-breathing engines, as the X-15 with a single rocket engine went 4,520 mph unofficially on 3 October 1967.

forgiving flyer both at high and low speeds, and it boasted the lowest accident rate of any single-engine aircraft in the Air Force. In those days several missiles had been stowed in its weapons bay, one of which might have a nuclear warhead, a spear to be hurled in some final, desperate war.[26]

When the winds of history shifted to a new direction, these interceptors no longer had a mission. Following their decommissioning, they had been stored at the Air Force depot at Davis-Monthan AFB in Tucson, Arizona. They were later removed from storage, modified for target service as unpiloted drones, redesignated QF-106s, and transferred to Tyndall AFB, Florida. Once a month one lucky individual was rewarded with a "hot" missile to demolish another 106.[27] Few of the airplanes remained. Down at Tyndall near Panama City, the last ones were parked, Cold War interceptors on the tarmac waiting to be used for target practice.

Could the F-106 be the towed airplane for the Eclipse project? Could KST negotiate an agreement to pull the old warrior on a rope? Another question intrigued KST engineers. Could the F-106 later be modified, outfitted with a rocket, and used as an operational launch vehicle?

On 22 May 1996, an Eclipse team representing KST, Dryden, and the AF Phillips Lab made the journey to Tyndall to look at the F-106s. It resembled a trip to a used car lot to kick the tires. Which of the remaining airplanes might serve the project? But a larger issue was not completely defined–corrosion. Years of sitting exposed to the salty air beneath the Florida sun had taken a toll on all aluminum parts in these airplanes.

KST had sent two veteran engineers as its representatives. The KST lead was Don Anctil, an engineer whose experience included work on numerous aircraft including the F-102, F-111, and C-5A as well as prototype design on the F-106. The other was Bill Drachslin, a designer who had worked on many different missiles and in his early years had been an Air Force maintenance crew chief on the F-86 in Korea. Anctil rubbed his grisly chin and stared at the Air Force faces across the table. His West Coast buddies had been taunting him. They snorted that Anctil might be on a mission to retrieve "tuna cans" and "hangar queens," industry terms for airplanes no longer suitable to fly.[28]

The initial briefing did not bode well. The commander spoke. He had orders to release a pair of F-106s, but he also had crash movies to show them first. The hopeful aspect of the F-106, he explained, was that the Air Force "had lost aircraft but no pilots to date." What was the problem? In essence, the problem was a 38-year-old airplane. The bad news was four crashes resulted because of failures in the aging landing gear. Cracks

---

[26] See, e.g., *Jane's All the World's Aircraft, 1964-65*, ed. John W. R. Taylor (New York: McGraw-Hill, 1964), p. 219; F. G. Swanborough with Peter M. Bowers, *United States Military Aircraft Since 1909* (London & New York: Putnam, 1963), pp. 154-155.

[27] The F-106 was variously called the F-106 interceptor and the Delta Dart. At Tyndall after the airplane was modified as an unpiloted vehicle, it was named the QF-106, and at NASA Dryden for the Eclipse project, it was named EXD-01 for Eclipse Experimental Demonstrator number 1. Both of these designations were local to very specific times and places. In conversation, Eclipse personnel who worked with the airplane during all of these stages often referred to the airplane simply as the F-106 or even 106. It is important to recognize these various names. But for the sake of simplicity, throughout this history, the airplane will usually be referred to as the F-106.

[28] Don Anctil, interview by author, 14 July 1999. Comments of KST reviewers.

had also been discovered in the wing spars of several aircraft, causing minor fuel leaks. The good news? Four times pilots ejected safely. But when the Eclipse team went outside to the steamy heat of the tarmac and hangar and talked to the crews, they received another message, one with a different emphasis.

Every airplane waiting in the rows had a personality, and the mechanics who worked on them knew it. They knew every inch of these aircraft. The maintenance crew had picked the two best they could find. They scurried about with records, logbooks, and grease-stained service manuals. Forger, Dryden's Tony Ginn (a young engineer assigned to the project) and KST's former crew chief, Bill Drachslin, climbed over the vehicles, peered inside, and took photographs. There were no hydraulic leaks, no fuel spills, no cracks in the control surfaces. In the briefing room, the message had been that the F-106 was marginal.

Out in the hangar, the emphasis was different. "Safe enough," said the mechanics. Age, of course, would remain a problem. For instance, most of the F-106's parts could not be replaced, simply because replacements were no longer available in warehouses. The fuel system was not maintainable if anything went wrong–it required 196 fuel valves. The tires were worn. The landing-gear support structure was suspect.

But as Forger, Ginn, Hampsten, Drachslin, and Anctil looked up beneath the airplanes the mechanics had picked for them, they exchanged smiles. These were flyable aircraft.

And the news got better. When Anctil attended subsequent meetings, he had the impression that at the Air Force's administrative level, the F-106s were almost an inconvenience. The command was looking forward with anticipation to an arrival of F-4s, a new generation of target drones. As Anctil tried to listen between the lines and plumb beneath polite phrases, his eyes grew wide. His pencil scribbled on the yellow pad, "If selected aircraft are modified beyond the normal F-106 envelope, Tyndall does not want them back under any conditions." His eyes grew even wider and he scribbled faster: ("Personal note: Tyndall does not want them back period!!")[29]

Another issue resolved as neatly. Mike Kelly had voiced the hope of acquiring two different models, the F-106A, the original single-seat interceptor, and the F-106B, a later modified two-seater. Kelly had public relations uses in mind for the second seat. He was a realist. He was not demanding or pressuring. Clearly, there were downsides with having two different vehicles to maintain. And the Air Force's "horror movies" raised liability issues. As the question was discussed in a tiny meeting room at Tyndall, Dryden's Tony Ginn jotted in his notebook, "Why risk two lives?"[30]

But Ginn did not have to voice his opinion. The Air Force's Dick Chase in a briefing pointed out that many significant differences existed between the models including different pilot training, maintenance procedures, aerodynamics, fuel systems, paperwork, official reporting, and correspondingly different simulation and test operations. A bonus of keeping two F-106As was that one could be "cannibalized" to supply the other with replacement parts that otherwise would be unavailable. Chase finished his presentation and sat down. The two-seat issue vanished.

In the months that followed, Forger, too, grew attached to the F-106. When asked

---

[29] Don Anctil, personal meeting notes.

[30] Tony Ginn, interview by author, 27 July 1999.

about it recently, he leaned back in his swivel chair in the Dryden pilots' office, balancing in midair. "It was," he declared, "a grand machine."

He especially liked the afterburner. The F-106 had one like none he had ever seen. Typically, when a pilot selects afterburner in modern engines, the fuel control meters in a small amount of additional fuel to spark plugs in the rear of the engine, which safely ignite the afterburner. Once it is lit, additional fuel is then available for full afterburner thrust. This gradual "light-off" results in a smooth acceleration. But when an F-106 pilot selects afterburner, a "bucket" of jet fuel is dumped into the hot exhaust for a sudden and dramatic torch ignition. There's a loud explosion, and the pilot slams into his seat from the dramatic increase in thrust.

"It was incredible. You'd select afterburner," remembered Forger, tilting forward in his chair, and then "there was a very pregnant pause. Finally, a big boom and off you go."

How robust was the F-106? At the start, Ed Skinner, a veteran engineer assigned by KST to examine the plane's maintenance records, smiled at the issue. He observed that although the aircraft seemed as ancient as some of KST's retirees, it was well maintained and still in great shape for the demanding tasks ahead.

Another Eclipse worker who became an F-106 admirer was Todd Peters, the youngest member of the team and an engineer who had recently graduated from college. After an early test to get some data on the F-106, Chief Engineer Al Bowers remembers walking away from the control room with Peters, who followed behind him in typical brash

fashion, making scathing remarks about working with ancient airplanes.

Bowers remembers a silence next, following behind him, and then a rustling of pages as Peters scanned the data. The young engineer's voice emerged again behind him, but much softer. There was a new note. It was awe.

"F-106 *rocks*," he said.[31]

In any event, the Eclipse project at last had an airplane to tow, a geriatric warplane, robust in its power but questionable, especially in a few unsettling aspects of its emergency and life-support systems. In the months ahead, heads would shake, camps of debate form, and several Dryden employees would find themselves called upon to make dramatic decisions. But when the group returned home on the airline from Panama City on 26 May 1996, questions had been answered, and a decision made.

F-106 was Eclipse.

* * *

Al Bowers became NASA's chief engineer on the Eclipse project that summer. At the time, real flight tests were only a proposal, but Forger must have glimpsed a chance. "I recommended Al," recalls Forger, "because he had both the engineering intelligence and also the passion to make it happen."[32] Bowers is a genial, dark-haired engineer in his mid-thirties who sometimes gets so excited about a flight validation that he has been known to leap up on a desktop in a technical meeting, shouting and pointing to his data printouts. But Dryden management had already spotted something in him far beyond a scientific cheerleader, appointing him as chief engineer on the prestigious High Angle-of-Attack Research

---

[31] Albion Bowers, interview by author, 25 June 1999.

[32] Mark Stucky, interview by author, 22 July 1999.

Vehicle (HARV) project. Behind his positive, upbeat approach was an engineer who could weigh positives and negatives and judge procedures and personnel assignments with a remarkable coolness and insight. He would serve the demands of Eclipse very well.

While management wrestled with funding issues, the team began to address the technology. In addition to Forger and Bowers, there was now Bob Baron who replaced Gary Trippensee as project manager. Bill Lokos came on board as lead structures engineer, responsible for ensuring that all modified and new structures were strong enough to ensure safety of flight; also, Jim Murray brought to the technical team his skills as an aerospace engineer and analyst; from simulations came Ken Norlin; Mark Collard served as operations engineer and the flight controller; and Joe Gera, a respected Hungarian-born engineer with half a century of experience in soaring, was called back out of retirement by Baron as the flight controls engineer. The team also included Tony Branco and Bill Clark, teamed as instrumentation engineers; Roy Dymott, systems engineer; and the newly-hired Debra Randall as test information engineer. Later they would be joined by aerial video photographers Lori Losey, Carla Thomas, and Jim Ross. For many naysayers about Eclipse as well as for NASA managers and potential investors for KST, it was videotapes rather than technical data that often proved the points Eclipse was trying to demonstrate.

From the start, there was debate. As the team began to plan flight-test procedures, the initial issue became "high tow," the traditional approach, versus "low tow." Traditional glider aircraft have large wing areas, resulting in large lift-to-drag ratios and correspondingly low takeoff speeds.

They take off before the tow aircraft and remain above them throughout flight, in what is called high tow. The F-106, on the other hand, has a much smaller lift-to-drag ratio and a correspondingly high takeoff speed of about 115 knots. To acquire a high-tow position would require the F-106 to traverse the C-141's wake turbulence from the initial low-tow takeoff position. This position would have been foreign to traditional glider experience.

There were fierce differences among team members. Jim Murray recalled the seemingly endless meetings. "Everyone's got an opinion," he smiled; "they're more readily available than ideas are." Every test program spawned differences, but again and again, Eclipse created a spectrum. "It *was* unusual how extreme the positions were," nodded Murray.[33]

Many of the differences were between people who had gliding experience and those who did not. If you had flown gliders or sailplanes or gone soaring, you had been at the end of a tow rope. If you had, towing was casual. It was matter of fact. Some felt simply that if it flew, it could be towed. Researchers with this background felt that there were almost no test issues. In their minds, the logical next step was simple flight. Gera sums up this viewpoint; he says, "It was a piece of cake."[34]

The gliding people tended to argue for the traditional high-tow position, apparently minimizing the risk involved in traversing the C-141's wake. And if gliding people grew emotional in debate, the response fed on the emotions experienced in thousands of hours of recreational flight on weekends. The clincher in the debate came from Jim Murray. His simulations demonstrated that the Eclipse

[33] James Murray interview.

[34] Joe Gera, interview by author, 16 June 1999.

had to fly low-tow. The sims indicated instabilities for the rope and the F-106 when flown high-tow, results which were in fact echoed, but more benignly, in later flight.[35]

The Eclipse project stayed aloft by more than technology efforts. There was also a social context. On 28 October 1996, KST scheduled a kick-off party. At Dryden people will tell you that in the genus and family of party animals, engineers have no place.

KST president Mike Kelly, of course, was an engineer. But Kelly, despite all the folklore and jokes about engineers and their poor socializing skills, did know how to throw a party. He arranged a splashy celebration for Eclipse in an old hangar at what had been Norton AFB in San Bernardino. There was food, drink, music, and the tables overflowed with more than six hundred people. Guests included two congressmen and NASA Administrator Daniel S. Goldin. Kelly had hoped to make an impact by displaying the F-106 at this party. After some debate, he had to settle for the C-141A and one of NASA's F-18s. When Dryden research pilot Ed Schneider departed the party early in the F-18, he swooped down over the merrymakers in a fly-by, evoking oohs and ahs.

Dan Goldin gave a speech. He described his vision of future NASA-commercial collaboration in space travel. He reiterated his mandate, "Better, Faster, Cheaper." And he gave a nod to NASA's collaborative partner in this effort and also to hundreds of other small, visionary start-up companies feverishly pursuing the dream of a breakthrough in low-cost access to space.

Coincidentally, that night a movie was being shot in another hangar nearby. On break, the movie stars and crew joined the crowd. If the movie people worked with the stuff of dreams, the Eclipse people did, too. As one engineer wandered through the crowd, he and his wife might turn and find themselves face to face with some starlet they recognized.

There were two sets of dream-makers in the crowd that night.

* * *

In the weeks that followed, the Eclipse team settled down to work. First it took a closer look at historical precedent, as Kelly himself did at the outset. As noted above, the earliest patent of the concept dated from 1919 and was awarded to the pioneering Anthony Fokker, but useful information was hard to come by. Because of restrictions on the use of powered aircraft in the Treaty of Versailles after World War I, the Germans did extensive experimentation with towed vehicles. But they did not create a body of theoretical literature. Nor had the sailplane and gliding fliers established validated numerical models. A few theoretical papers had found their way into journals. Murray described the flight-test information on towing as "largely qualitative and anecdotal."[36] If the Dryden Eclipse team members needed data, they would have to do the tests themselves.

* * *

Of all the agencies KST negotiated with– and they were legion (Mike Gallo, KST vice president for marketing and sales, once estimated that he had negotiated with more than 33 federal units and sub-units in managing Eclipse)–Tracor Flight Systems, Inc., the F-106 maintenance contractor, seemed to present the least likelihood of creating problems. This was

---

[35] Bowers interview.

[36] Murray interview.

QF-106 aircraft in flight during February 1997 before the tethered flights began. (NASA photo EC97 43932-12 by Jim Ross)

a commercial firm. It had a hangar, first-rate technicians to service the F-106, and the original drawings from the manufacturer. The NASA pilot was to fly two Air Force F-106s used by the Eclipse project and just park them at the Tracor facility in Mojave, California. The support was expected to involve a simple money transaction. There would be none of the paperwork and serpentine federal-government procedures involved in interagency transactions. But there were glitches. Baron and Forger found themselves frustrated and stalled when they tried to arrange to fly the airplane. Because the planes still belonged to the Air Force, that service's local representative was required to enforce every regulation. No one at Dryden enjoys remembering those days.

Behind the scenes at Tracor, however, events were occurring over which KST had no control. A recent restructuring had placed responsibilities for management of the Mojave work at Tracor headquarters in Austin, Texas. The company also had been fortunate to win a lucrative contract with the Boeing Company at the Boeing facility in Palmdale, California. Tracor priorities, therefore, had shifted dramatically since the initial arrangements with KST. Consequently, disputes began to arise between KST and Tracor over work performance and compensation. It appeared to KST that Tracor was charging more and doing less.[37] "My guess," said Bob Baron, "was they put such a high price on it because they didn't want the business."[38] Clearly, Tracor had its hands full with much larger projects crucial to its own future. KST also was driven by profit. But in the mega-budget world of aerospace, it could get driven out by profit, too.

---

[37] This section based on KST comments on the original draft of this monograph.

[38] Robert Baron, interview by author, 11 June 1999.

At this point, Dryden would cross the Rubicon. The decision would be made in May of 1997 for the Air Force to transfer the F-106s to NASA, which would house them, service them, modify and instrument them in a Dryden hangar. (Because there was only one government agency involved in flight approval, the business of flight research was simplified.) And with these arrangements new responsibilities for flight safety began falling into place–not without debate.

The Air Force C-141 team had it much simpler. The Air Force owned and operated the C-141A. In fact, the 418th Flight Test Squadron had a C-141A in its hangar at Edwards AFB. The 418th had qualified Starlifter maintenance crews, and it would supply the pilots, the engineers, the technicians, and the ground and flight crews, albeit on a non-interference basis. In other words, Eclipse's work would get done, but without any priority. As Carol Reukauf, who replaced Bob Baron as project manager, later noted, when you looked on the Air Force priority list, there Eclipse was, on the bottom, number 17.[39]

The 418th assigned Capt. Stuart Farmer as its project pilot. Farmer, a dark-haired young man with an affable manner who revealed a sharp interest in technical issues in the months to come, was a "new kid on the block," just as Forger had been. For several weeks, Farmer had been sitting at his pilot's desk without any major projects to work on. When he was called into a meeting and asked to respond to very skeptical questions about towed flight, Farmer gave the concept thumbs up. He later admitted he was not sure of the aerodynamics issues. He just

wanted to fly. By the late date at which Eclipse actually flew, Farmer would have five other Air Force projects on his hands, and non-interference would become an issue. But initially, his enthusiasm helped keep the project alive.[40]

The 418th assigned Mark Watson as its project manager, replacing Bob Plested who had guided Eclipse through the paperwork of transferring the F-106s to NASA. Watson is a heavy-set young man with a shrewd ability for making things happen. Co-pilot Kelly Latimer came to the project fresh from the U.S. Air Force's Test Pilot School. A slender young woman with reddish hair in a Joan-of-Arc cut and a sense of humor, she also qualifies as one of many Eclipse landmarks: when Latimer flew in the right seat on four of the Eclipse flights and the left seat on two missions,[41] she became the first woman ever known to fly as a pilot on a NASA Dryden flight research mission.

Other Air Force crew and personnel assigned included Morgan LeVake, operations engineer; Bob Wilson, the lieutenant colonel who oversaw safety; Roy Surovec, the deputy Air Force project manager; Senior Master Sergeant John Stahl, the chief flight engineer; Art Tecson, who handled instrumentation; the scanner, Sergeant Dana Brink, source of some brilliant unofficial aerial photography; and Sergeant Ken Drucker, the loadmaster, assigned vulnerable duty at the end of the rope.

For the Air Force, answering operations questions for the C-141A was simply a matter of looking in the regulations. But for Dryden–and to the dismay of the

[39] Carol Reukauf, interview by author, 11 August 1999.

[40] Farmer interview.

[41] Her two flights as pilot rather than co-pilot were flights 8 and 9 (tethered flights 4 and 5), 28 January and 5 February 1998. Daily/Initial Flight Test Reports, C-141A, Flights F-5 through F-10, 20 Dec. 97, 21 Jan. 98, 23 Jan. 98, 28 Jan. 98, 5 Feb. 98, and 6 Feb. 98 respectively (see documents 16, 24, 32, and 44). Incidentally, Latimer was a major.

commercially-driven Kelly–in the business of aerotow, it was a matter of making engineering science. As Dryden increased its presence on the project, two new goals were added to the experiments: one, the establishment of safe operating procedures, a Dryden hallmark over the years, and two, the discovery of new technical information, Dryden's primary purpose as a flight research organization.

As the project gained status, Eclipse flight began to seem remotely possible. Forger and Gera both argued that flight safety was a non-issue, but Dryden scheduled batteries of ground tests and flight simulations to make sure.[42]

During the summer of 1996 the Dryden pilots took cautious note. Several thought that the greatest risks attended the takeoff; there were scenarios of rope break or accidental release, slacks and snarls about airplane gear. The hazard scenarios were many. Joe Wilson remembered a conversation with Gordon Fullerton, ex-astronaut, crackerjack pilot, and a shrewd, practical thinker about flight issues. Wilson recalled Fullerton cocking his head, pointing out that there was no forgiving altitude. In the simulator, Forger had been doing inadvertent releases at 10,000 feet–at which altitude, if something went wrong, he had some time to plan and do something–but if something happened on the Eclipse takeoff, Forger only had his reflexes.[43]

If something went wrong with Eclipse at a low altitude, it was going to go wrong fast.

Dryden Chief Engineer (and former Chief Research Pilot) Bill Dana also questioned the safety of Eclipse. He explained that he personally had a sense Eclipse flights could be done but that as chairman of the Airworthiness and Flight Safety Review Board, his job was to raise safety questions. "I was the devil's advocate," he explained.[44]

Names

If you ask Mike Kelly where the name Eclipse came from, he doesn't blink or hesitate. He recalls that he and Mike Gallo dreamed it up in their conference room one afternoon. What does the name stand for? He acknowledges there is no significance–it's a name with a "feel," easy to broach in a meeting, lofty sounding, a bit of verbal flare short on the denotative aspect of language. In blunt fact, there is no eclipse in the Eclipse project.

But names can decide destinies. If you pick the right name, Dryden engineers say, it helps when you appeal for budget or support–especially if you find yourself in competition with another project as worthy as your own. And some engineers say that the wrong name, an unusually clumsy one, can do harm. At NASA Dryden, the engineers understood the importance of names to bureaucratic approvals, and over at KST, they also understood the importance of a name when approaching investors or a bank.

A second name appeared later. It was an unofficial name. To this date, no one claims to be its coiner. It first appeared in public one day when Forger, climbing into the cockpit dressed in pilot's suit, test point cards clipped to his knee pad, looked down. He saw a rough inscription hand-painted on the side of the F-106.

[42] Mark Stucky, interview by author, 22 July 1999.

[43] Joe Wilson, interview by author, 28 June 1999; Gordon Fullerton, interview by author, 26 July 1999.

[44] Bill Dana, interview by author, 26 July 1999.

Certainly the gleam of humor blessed the name, some inscription dreamed up perhaps during a stop at a desert saloon on the drive home, but it also fed on the dismay of expert pilots back at Dryden concerning Eclipse. It read:

DOPE ON A ROPE

Daryl Townsend had been present that day. The crew chief remembered peeking around the maintenance truck. Forger was new. What would he do? If he was a by-the-rules sort of guy, a storm would follow. Dryden was a flight research center, and without expert research pilots, it could not do its business. Thus, although they were often the butts of jokes, pilots also had formidable clout, which they could wield.

There was no storm. The new pilot paused. Townsend describes a smile perhaps, a subtle nod of the head. Subtle enough that Townsend had to ask Forger later, was he *sure* he didn't mind? Forger said it was OK.

The crew didn't scrub the name off.[45]

Subsystems and Worry

One mechanism needed for Eclipse was called by the technicians the "knuckle," a hunk of metal, three pounds or more, much larger than a human knuckle in fact, larger than a heavyweight's fist, a nasty bit of hardware in some events to come but created for elegant purposes. It was crucial.

If the sole project intent were to pull an airplane, the knuckle could be omitted. But if technical data was needed or if the pilot needed real-time information on what was happening to the rope in flight– and in fact he did–this knuckle was a necessity. This universal joint attached to

the release ironwork, gave the rope free play, and instrumented both azimuth and elevation angles of the rope.

The Dryden engineers moved swiftly to analysis and testing. Much of the analysis concerned the rope. "One assumption we made early on was that the lift and drag of towrope is negligible," explained Bowers, "but that was an invalid assumption."[46] If that was not surmised, much else was. As soon as they decided whether they would operate in high tow or low, the engineers could start looking for solutions. It was a given that the rope would attach to the rear of the C-141A. In low tow, the rope would attach to the top of the F-106.

But where some glider enthusiasts may have assumed the rope had to attach near the center of gravity (CG) of the F-106, the technical requirements for the Eclipse airplane were different. In fact, the relationship of the distance of the tow attachment to the CG as compared to the distance of the control surfaces to the CG was the exact opposite of the arrangement that occurred on a conventional sailplane. A sailplane has the rope attach close to the CG while the control surfaces (elevator, rudder, and ailerons) are some distance away from the CG. This means the tow forces can easily be countered by the aerodynamic control forces. On the F-106, the tow attachment was in front of the canopy while the CG was located many feet farther back in the center of the airplane, much nearer to the control surfaces. This meant the potential existed for tow forces that could exceed the pilot's ability to counter them.

Once the engineers had a plan for takeoff configurations, they could make other decisions. What would be the rope length? How much weight would the rope bear? What stresses did it have to

---

[45] Daryl Townsend, interview by author, 25 June 1999.

[46] Albion Bowers, interview by author, 8 June 1999.

F-106 tow cable attachment and release mechanism for the Eclipse program. (NASA photo EC97 44233-5 by Tony Landis)

endure? These and other difficult questions required answers.

Kelly's original plan had been to reuse tow rope. To be sure, the rope came in expensive at $9.30 a foot. Perhaps KST grew impatient with Dryden's approach to decisions about the rope. Or perhaps it was a generational thing–the majority of KST's employees were gray-haired semi-retirees who came of age working on aircraft and ballistic-missile projects back in the 1950s. (Their employment as part-time workers was one of Kelly's efficiencies.) In personal remarks in interviews, younger Eclipse team members often brought up generational remarks; they looked across an age gap at the older engineers, sometimes with fascination, sometimes with dismay, and occasionally with humble respect. One youthful engineer described the KST retiree-engineers as the kick-the-tires-and-go-fly generation.[47]

---

[47] Phil Starbuck, interview by author, 29 July 1999.

And they used that approach with Vectran® rope abrasion tests. With genuine zest and enthusiasm, two KST engineers, Archie Vickers and Bill Williams (system engineering manager and test manager, respectively, for KST), describe an impromptu test of cable reusability. They took a length of Vectran® to a Hemet Valley airfield where they used it all day towing gliders. They beat it on concrete. They beat it on gravel. Breathless, they beat it finally on dirt and tossed it in a box where it rumbled with sand, dirt, and rock on the drive home. Although they noticed slight damage, they came to the conclusion that the rope was reusable. The rope was tough.

Early on, KST had investigated a cable spool to reuse the rope. After such reuse, would the rope still be as strong? Would it degrade or carry over unsettling memories (energies) from the coiling? Dryden pointed out it would be less expensive and speed up the schedule simply to buy multiple ropes and use each of them only once, thereby eliminating a good deal of fabrication and testing. Dryden's agreement to purchase the additional ropes made the decision easy for KST.

As the rope questions were slowly answered, the subsystem work moved along. Tony Ginn had the early inspiration to convert an Air Force parachute qualification pallet to the uses of airplane towing. The pallet was already flight-qualified and designed to be attached to the floor at the rear of the C-141A. This concept saved months of development, design, fabrication, and testing. The pallet came complete with a guillotine designed to cut the nylon straps used to attach the heavy loads to the extraction chutes. Rope release devices constituted a crucial safety issue and here was an unplanned blessing. But when they loosed the spring-load force of the guillotine blade, it failed. It would not cut the tough Vectran® rope. The solution was to attach the rope with a three-pin connector designed by Dryden contract-employee Roy Dymott to a nylon strap, a substance the guillotine could slice. If that should fail, the loadmaster might cut the nylon strap with a hand knife (a device which, to outsider eyes, resembled a small ax).

The device for releasing the F-106 from the rope also proved an unforeseen gift. When operations engineer Bill Albrecht, who had long been associated with the B-52, attended a planning meeting for Eclipse, he asked, why not use B-52 parachute release hardware, a device that resembled an iron jaw?[48] This was qualified hardware, in regular use, in Air Force stock, and would more than carry the load. Forger could activate the release jaw electrically, and in case of malfunction, he had a mechanical backup.

The emergency release device for the F-106 was the frangible link, or "weak link" as it came to be called. The frangible link–a safety mechanism–would break before the rope or nylon ever broke. Although it was designed for emergency release, on later flights the Eclipse team started breaking the frangible link to release from tow because it kept the instrumented knuckle assembly attached to the F-106's release mechanism where it could readily be used again. When the team initially used the release in the configuration designed for the first flight, the knuckle was on the end of the 1,000-foot rope still attached at the other end to the Starlifter. It whipped so wildly in the hurricane-force winds that the frangible link snapped and the knuckle was lost in the desert.

---

[48] Bill Albrecht, interview by author, 17 June 1999. According to Al Bowers, the idea arose earlier among Collard, Forger, and himself, but it could not be implemented without Albrecht's OK. Bowers' comments on a draft of this monograph.

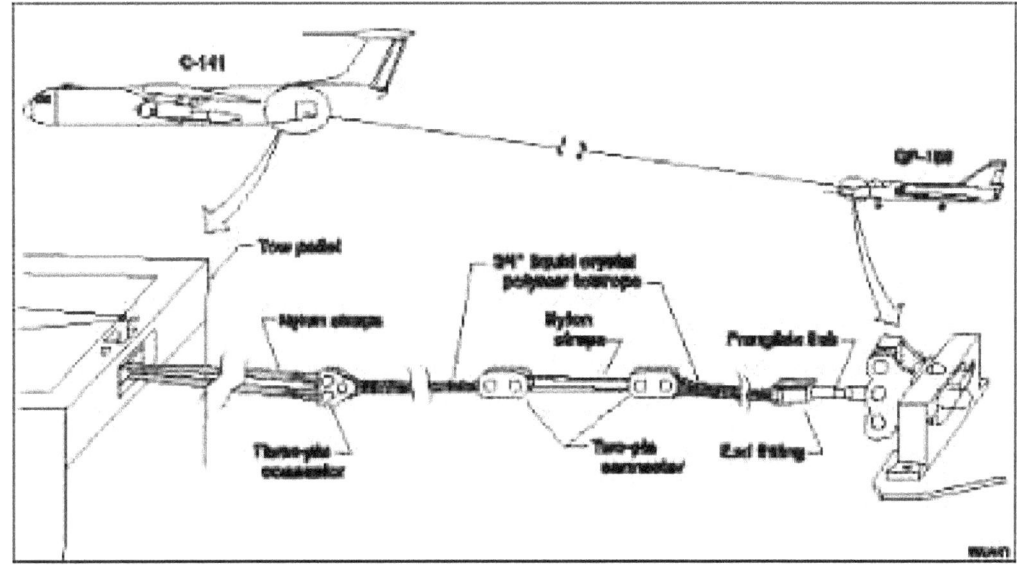

Figure 2a. Schematic drawing of the initial tow-train configuration. (Design 980441 by the Dryden Graphics Office)

KST had designed the basic frangible link. Its initial plan had been to couple it with off-the-shelf load cells from a commercial source. At Dryden, however, Bill Lokos redesigned the link; it was a nifty solution that eliminated the need for a separate load cell on the C-141A (tow-train) end of the assembly. To accomplish this, Lokos incorporated an integrated load measurement feature using two full-strain-gage tension bridges installed in the link itself, and also made other modifications, including changes in the alloy to ensure proper hardness through-out and changes in the neck diameter of the link (on the basis of extensive tension-failure testing). With these modifications, Bill was confident the link would break at the predicted load. The concern on the issue of obtaining a consistent breaking load continued. The solution was machine-shop fabrication and calibration of the links, each of which was to be used only once. To ensure consistency, all ten of the links to be used in the ten planned flight tests were made from the same lot of steel bar stock that supplied the links used in lab testing.

Along the way the team divided sharply into two camps. The strength of the weak link had to be decided upon relatively early in the design phase because its strength, by definition, set the maximum loading the F-106 could be subjected to.

Figure 2b. Schematic drawing of the simplified tow-train configuration. (Design 980497 by the Dryden Graphics Office)

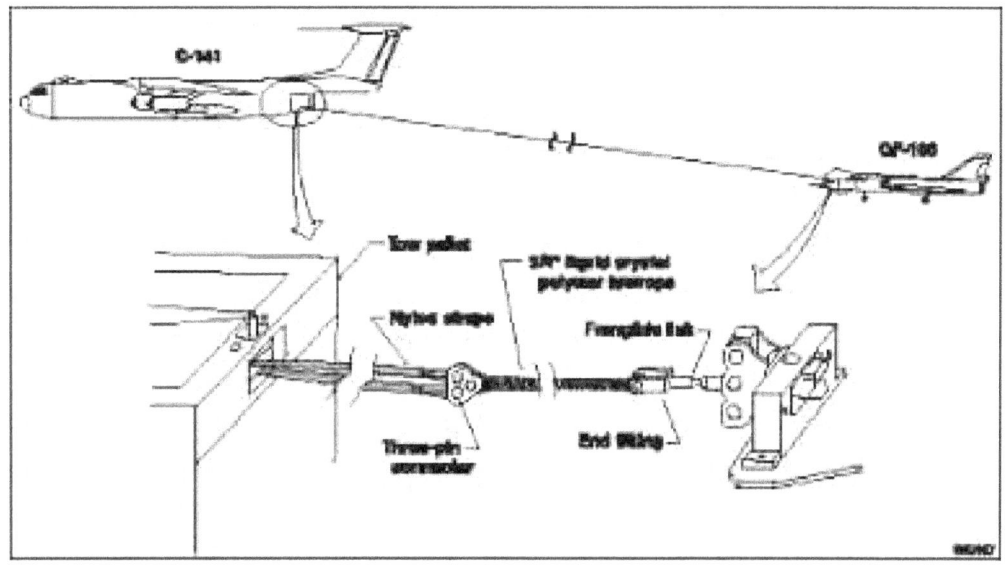

The stronger the value of the weak link, the greater the potential loading of the fuselage and the greater the "beef-up" required to the fuselage. The Federal Aviation Administration's regulations for gliders or sailplanes stipulated that the link must break at a maximum force of 80 percent of the weight of the glider being towed. If this criterion were applied to the F-106, the breaking strength would be approximately 24,000 pounds. Although the Eclipse tests were not subject to FAA regulations, this figure was a valuable reference point for design of the frangible link for the F-106 tow-testing.[49]

There were those who were advocates of "strong" weak links and those who advocated "weak" weak links. The "strong"-weak-link group was concerned primarily about the hazards of a low-altitude, inadvertent link breakage and felt the F-106 would crash into the desert if the weak link broke during the critical takeoff phase. The "weak"-weak-link group, of which Forger was a vocal member, was more worried about the stability-and-control issues under tow and wanted the weak link to break before the airplane could go out of control on tow. For this group's argument to prevail, its members first had to demonstrate that the F-106 could power up quickly enough to fly out of a low-altitude, emergency release before disaster ensued.

Forger's claim that he could fly the F-106 out of a low-altitude, inadvertent release was eventually accepted. Using the newly instrumented Eclipse aircraft, he demonstrated landing approaches in which he swooped down with the engine stabilized at idle, the landing gear down,

and the speed brakes fully deployed–the worst-case drag situation. He held the aircraft inches off the runway as the airspeed bled down to 150 knots, a full fifteen knots less than the planned tow takeoff speed. This slow speed simulated a rope break at the most critical time, including several seconds for pilot reaction. Forger then selected "military power"[50] and retracted the speed brakes. The venerable J75 engine took six seconds to spool up, during which time the F-106 slowed precariously, but Forger was always able to maintain control until usable thrust was regained. The test was repeated numerous times, the data strips demonstrating conclusively that the F-106 had the flying qualities and engine response to fly out of any threatening situation from the moment the aircraft left the runway.

Ultimately, the "weak" value of 24,000 pounds was accepted for the weak link. On the eve of the first flight, there still remained a number of team members who thought the link should have been significantly stronger.[51]

Another problem was that although the C-141A had an off-the-shelf tow rope attachment available for a tow assembly, the F-106 did not. The KST engineers remedied this by providing a weldment apparatus that was riveted to the nose of the F-106. It was black, a bizarre object. Because of its shape, the crews called it The Bathtub. Like other new structures, it had to be tested by structures engineer Bill Lokos.

Meanwhile over at KST, Wes Robinson led his engineers in shepherding the rope through breaking tests subcontracted to a laboratory in Los Angeles. "When the

---

[49] Based on KST and Bill Lokos' comments on the original draft of this monograph.

[50] The term "military power" refers to the use of maximum power without use of the afterburner. It is differentiated from "maximum power," which includes the use of the afterburner.

[51] This section of the narrative is heavily indebted to editorial notes from Mark Stucky, 16 September 1999.

rope was at last close to failing," remembers Robinson, "it got hot and it would weld and I remember the smell, that burnt plastic sort of smell."[52] When the rope finally snapped, the sound resembled a small cannon being discharged.

* * *

Eclipse had started with Gary Trippensee as project manager and had transitioned to Bob Baron and now changed again. Carol Reukauf, a diminutive woman in her forties, came aboard as project manager in April of 1997. Reukauf tended to be casual in manner, but behind the informal appearance was a woman with remarkable organizational abilities and a shrewd ability to deal with groups of people. She came aboard just as the mechanical assembly of the fixtures on the F-106 converted it to its tow (EXD-01) configuration and the ground testing began. This was also when a series of safety review meetings appeared on the horizon, a few of them viewed as threatening by the team. "There were procedures and papers to be filed," says Bowers, "and we knew she would be good at it."[53]

In the flight reports, however, Reukauf wrote in a style very different from that used in typical NASA reports. Her language seemed to come from the world of self-improvement and group support. For example, her last report states, "I advise everyone to reflect on their Eclipse experience, take the personal lessons that you learned and apply [them] to your future endeavors."[54]

Her upbeat comments in these reports were, in a sense, directives. They were not a threat. But in retrospect, they firmly pointed many people in the same direction at a point when the multi-partner effort seemed in some weeks about to collapse. "It was important to stay on a positive note," she says, "because you don't need any negative notes when you are trying to get the project done in a rush."[55]

She was also famous for extended meetings, although she insists they never lasted more than one and a half hours; they happened every Tuesday morning in the lakebed conference room, a meeting area that looked out on the runway. The primary Eclipse members were required to come, and her insistence kept everyone focused, every unit and agency in the loop. If you ask today, many Eclipse members report an unusual sense of involvement and fun with the unruly project. Ken Drucker of the Air Force, for instance, testifies, "It was the highlight of my career."[56]

To the dismay of some, Reukauf involved as many members as she could in debate on issues that were related to safety, instead of deferring to expert opinion only. Bud Howell, KST representative at the weekly meetings, noted that Carol's insistence upon hearing all sides of an issue had a very positive effect on team morale. Forger, on the other hand, recalls lengthy discussions spent "ferreting out the ridiculous."[57] Reukauf's response? "It was good for Forger," she smiles. "We needed everyone to consider the ramifications of decisions on this complex project." She

[52] Wes Robinson, interview by author, 30 July 1999.

[53] Albion Bowers, interview by author, 10 August 1999.

[54] Project Manager's Comments, Eclipse EXD-01 Flight 10, 6 Feb. 1998, Eclipse Flight Report (see document 41).

[55] Reukauf interview.

[56] Drucker interview; Reukauf's comments on a draft of this monograph.

[57] KST comment; Mark Stucky, interview by author, 22 July 1999.

also emphasizes that she applied this approach "for the specific reason that I was concerned that the stronger, more articulate members of the team tended to express their views to the exclusion of those who differed with them." She saw the potential for damage to a team whose members already had sometimes contradictory agendas.[58]

Reukauf also made a point at the end of each meeting to ask for a comment from each member. "Jim Murray would sit silently throughout the meeting," she recalled, and when she called on the brilliant engineer at the end, "he would bring up a point nobody had thought of– and usually, he was right."[59] In hindsight, her approach created a unified team.

One of the great debates raging was whether the rope's oscillations might develop some pitching motion or unstable energy. The antithesis was the straight-line rope illustrated on the early report covers. Joe Gera defended straight-rope theory. "You can't push a rope," he said. Jim Murray argued differently. He suggested there might be a bungee effect. "What," he asked, "if the rope goes *boing-boing*?"[60]

Later, Murray and Gera decided to put the question to an unauthorized test. It was a good-natured jaunt–and also a secret as far as management was concerned. The two signed out for a day of leave (vacation), borrowed some load instruments from the lab, and set off to do the experiment on their own. They found a glider-towing company with an owner cynical but willing to pull their rented glider on an instrumented rope behind his tow airplane so

they could gather data. Space technology? Uh-hum. When the two returned to Dryden the next day, word had already reached project management. Reukauf spoke with each of them immediately. She came on tough, but curiously, Murray remembered, "It was very much like a mother scolding a child."[61] It was a tone that commanded, and she halted a growing Eclipse tendency to take legal risks on this high-visibility project so casually.

Reukauf herself in a few days found an authorized way for them to continue these valuable tests and still deal with liability issues. In fact, she found a way to use a federal government credit card and adhere to regulations about use of government equipment. This permitted Murray, Forger, and Gera to gather more experimental data in an unconventional way.

One great fear of skeptics was that somehow the wake turbulence of the C-141A would upset the F-106. In one test in the fall of 1996, the experimenters put smoke generators on the wings of the Starlifter and flew to see what patterns were traced in the sky. Forger took a leading role in actual flight tests addressing the issue. There were several factors. One was downwash, the streaming of air off the transport's wing, a disturbance that later Forger described as no more unsettling than driving a car on a gravelly road. But the big concern was vortices, severe air disturbances coming from each wingtip of the mammoth transport. The vortices proved to be small tornadoes which, as they moved away from their source and increased in size, for some distance at least also increased as hazards.

In the spring of 1997, Forger flew an F-18 into the wake of the C-141A. He flew in

---

[58] Comments by Reukauf on a draft of this study, September 1999.

[59] Reukauf interview and corrections in her review of a draft of this study.

[60] Murray interview.

[61] Ibid.

The aft end of the C-141A tow aircraft. (NASA photo EC98 44392-1 by Jim Ross)

near the transport's tail. He would take stabs at the vortices with his wing tip and every time he did, the F-18 rolled off. Even at a distance, team members could see a vortex. "Sometimes it would mix with the exhaust blowing, and in the glint of the sun," remembered Mark Collard, "you could see it was tubular. I could see it. He could, too, at times."[62]

How big across was the vortex when encountered a thousand feet behind the transport? "About as wide across as a volleyball," grinned Forger. "It was a non-issue."[63] So there he was up in the sky, playing volleyball with violent air. Later in the summer, he flew the F-106 behind the Starlifter in similar tests. There were no problems for Eclipse.

One regulation did, however, become an issue. The engineers had air-speed and altitude windows they wanted to investigate to validate the research simulation. The hunch was that an airspeed around 300 knots would provide ideal towing

conditions. If the petal doors were open in the tail, however, regulations required the C-141A to fly at less than 200 knots. The Eclipse team asked: if the petal doors were removed, did that speed restriction still exist? The petal doors provided no structural stability. Obviously, the restriction came from a concern with unstable dynamics on the opened doors.

Lockheed, the manufacturer of the C-141, had performed dynamic analyses for flight with the doors open because users needed to know the maximum speed for pallet air drops, which required, of course, open doors. Authorization to fly at a greater speed with the doors either open or removed would require further analysis by Lockheed. Reukauf remembered that the Eclipse team resigned itself to the limit because there was "no time [or budget] for a new stability analysis." But to this day, Ken Drucker, the Air Force loadmaster, regrets that he did not intervene in time with informal advice to get the team past the barrier.[64]

[62] Mark Collard, interview by author, 18 June 1999.

[63] Stucky interviews.

[64] Drucker interview; comments of Reukauf on a draft of this study.

On many other occasions, Drucker and Watson did in fact help the Eclipse team navigate around Air Force regulations. But the speed limit remained at 200 knots.

* * *

Dryden and the AFFTC may have shared the same runway, but they came from two different cultures. Often parties to both agencies would have moments of culture shock. Once Watson remembers departing one of the lengthy Eclipse meetings accompanied by Lieutenant Colonel Bob Wilson. Wilson shook his head slowly at what he had just been hearing, astonished at the intense interest of the NASA people in issues that struck him as purely theoretical.[65]

KST felt these cultural differences, too. Late in the summer of 1997, KST project manager Bob Keltner paid a visit to Dryden. He had worked on the Atlas missile earlier in his life and later spent decades at TRW. He got out of his car in the sweltering heat of the Dryden parking lot with some trepidation. He was about to present a list of grievances to Carol Reukauf. It was a curious document roughly printed in capital letters by hand. The title was "PROGRAM DELAY RESPONSIBILITY." He noted quite a number of these responsibilities and attributed a few of them to KST. He next had penned a section entitled "ACTS OF GOD," which left, of course, "ACTS OF NASA."[66]

There were many acts of NASA, a substantial number of them concerning Dryden's level of safety preparation and Dryden's commitment to generating data. It was another clash of cultures, really. And any slips in the schedule related to government regulation or a need for

additional safety factors or simply curiosity about some interesting data and the time taken to pursue it, all added up to expenses for KST–and new trips to the investors to keep the project floating.

Later Keltner told his KST associates about Reukauf's reaction. She sat a moment in silence after reading the pages, her hands folded on the table, then started shaking her head back and forth.

"You know, I am really disappointed in you," she said. It was couched in a sympathetic tone, but he could sense the iron in her, too. "No question," Keltner told his colleagues, "she was one angry lady."[67]

But curiously, the conference did seem to clear the air. Some of the issues were simply non-resolvables. But Keltner noticed that now at the Tuesday meetings when the Dryden data-gathers and analyzers threatened to stampede, she appealed to them to consider KST. She reined them in.

* * *

Many safety issues had to be resolved. One concern was the cockpit canopy. During a Configuration Control Board meeting someone asked, what if the stress on the F-106 fuselage bent the fuselage to the point the canopy could not be jettisoned? In an emergency scenario, it would entail disaster because the canopy had to be jettisoned from the aircraft before the ejection seat would fire. "This was another question that the project team judged to be a non-issue," explains Reukauf. "But nonetheless, regard to flight safety dictated a responsible pursuit of the real answer." The engineers moved quickly to gauge the risk. They found a replacement canopy and Dryden's structural testing lab under Bill Lokos'

---

[65] Watson interview.

[66] Robert Keltner, private papers.

[67] Robert Keltner, interview by author, 30 July 1999; comments of Keltner on a draft of this study.

Canopy stiffness
test setup.
(NASA photo
EC97 44303-01
by Tony Landis)

direction loaded the fuselage with shot bags and stressed the fuselage with loads which would be experienced in towing. It was not elegant, a rough sort of test. But the rough, reassuring answer was that the pyrotechnics could still blast the canopy free.

One of the many operational requirements identified in the initial KST test plan was to provide a cockpit display of rope tension for the pilot. This display was the work of Phil Starbuck, a brilliant young engineer (formerly of KST). It consisted of a horizontal row of lights that would change colors, ending in red as the rope load reached prescribed limits. In the millisecond scannings and judgments required at take-off, the monitor was a necessity.

Someone also had to weave an attachment loop in the rope. This was no small

task, because the splice had to retain the full strength of the virgin rope. The assignment eventually went to Dryden life-support technician Kelly Snapp. "Because I spent some time in the Navy," grinned Snapp when you ask him why.[68] He was adept at splicing a loop in the Vectran® lines. It was a skill, and if you thought the task simple, when you watched what Snapp had to do, it seemed a difficult and tedious trick.

To be sure, it was a task that might take an outsider half a week, but the ex-Navy technician could do it–and without damaging the rope, which was crucial–in perhaps half an hour. "He was quick," recalled crew chief Daryl Townsend in admiration.[69] And Vectran® did not cooperate when Snapp went to cut it. For all the worry about the vulnerability of the rope, he could wear out six to eight

---

[68] Kelly Snapp, interview by author, 25 June 1999.

[69] Townsend interview.

blades or dull one sharp hacksaw trying to cut it. Although Snapp never damaged the rope, his razors did slip his own way and on several occasions, the rope ascended into the skies with his loop *and* his blood stains on it.

Bill Lokos ground-tested six loops spliced for the experiment. Every time the rope failed, not the loop. And during the flight tests, the loops always held.

One of the paradoxes of the Eclipse project was that such a small project generated a number of landmarks. One of these developed when Al Bowers and Ken Norlin devised a simulation for the C-141A. They first modified the existing F-18 simulator at Dryden to represent the F-106. The Air Force did not have a C-141A sim available at Edwards, and as a result, NASA tests were producing useful results for the F-106 but none for the Starlifter. Early on, because the transport was so much heavier than the F-106, the Dryden engineers had modeled it in simulation simply as what they called an infinite mass. Bowers then addressed the need for a C-141 sim. Once this was done, the engineers set up three simulations–of the C-141, the rope, and the F-106–to operate together in real time (simultaneously) with data exchanged among them based on the dynamics of the simulated tow rope between the F-106 sim and the C-141 sim. The researchers actually set up the sims in separate rooms with radio communication between them and a control-room unit. It provided valuable rehearsal for the complexities which were only beginning to be recognized. It was groundbreaking.

Another landmark was the engineers' clever GPS contrivance. GPS, Global Positioning System, is a technology that uses satellite information to calculate exact location and rate of change–for instance, an airplane's geographic location and speed in flight. Most previous GPS uses consisted of linking one moving unit to a stable reference point. The Eclipse engineers scored high marks when they used GPS to chart in real time distances between two moving units, the tow airplane and the towed F-106.

* * *

On an August afternoon, Mark Collard sat in his office cubicle and stared at a paper. He hesitated. This was a moment when a person might take a long, deep breath before signing. The memo had just issued from his printer. A space waited at the bottom of the page for his signature. The whole business had to do with the pyros, the tiny units of explosive hardware, the only devices which would enable the Eclipse pilot to eject if there were a catastrophe. To no avail the engineers and support crew had searched for replacements, and none were to be had. The pyros on the F-106 were long past their expiration dates. This document would approve an extension.

He searched for a pen, found one–a federal-issue ballpoint. Such extensions were not unusual in flight research at Dryden, but if a problem arose in flight, the pyros had to work for Forger to eject successfully.

It was not a reckless moment for Collard. But the step raised questions. How much confidence do you have in this project? How deeply do you believe the presentations made in your own safety briefings? Are you sure go-fly hysteria has not taken over? Are you certain the momentum of fifty people working on this project for a year and a half is not the energy fueling your decision?

He signed his name. And he sent it to Tom McMurtry, Director of Flight Operations, who surely had his own internal debate before he signed.[70]

---

[70] People interviewed for this history voiced two different viewpoints about the extension for use of the pyros. One person argued, "The fact that we needed senior management to approve the extension means we were thorough." Another view was that Collard's signature, at least, could have had career-ending ramifications.

Extensions were an issue with the Air Force's aging warrior, too, and the C-141A had run out of time. There was a PDM awaiting the Starlifter that could not be avoided if it were to continue to fly. A PDM, Programmed Depot Maintenance, is a four-year cycle of attention that must be paid to Air Force airplanes. "This is *serious* maintenance," explained Bob Plested, the first Air Force project manager for Eclipse. "They take the airplane down to Warner-Robbins AFB and basically take it apart and put it back together again." The cost of a PDM weighed in at nearly three million dollars. And there was no Eclipse budget to come to the rescue. Because there were no other paying customers for the C-141A, the Air Force had decided to retire it for good.

"We had gotten another six months," Plested continued, "but there were no more extensions. It's what you call a drop-dead date. The first six-month extension is pretty much paperwork. But the second is bought more dearly." When all the Eclipse instruments and modifications were stripped out of the C-141A, wherever the airplane ended up on 18 February 1998 was going to be its final resting place.[71]

There were many safety reviews of the Eclipse project. Their number was extraordinary. Some personal comments were quite intense. One individual sent a memo concerning Eclipse flights that stated, "There have always been projects where people were willing to go out and kill someone, and this is one of them."

Dryden Director of Aeronautics Research and Technology Dwain Deets remembers that at NASA Headquarters in Washington, DC–where his job took him frequently–three or four times a week someone would come up and ask about this tiny project. "Would you give me a briefing?" he was asked. Deets notes that this modulated into a different question, "Are you sure of what you're doing?"[72] The informal reviews numbered in the hundreds.

Ken Drucker, the Air Force sergeant who was in charge as loadmaster at the rear of the tow plane, faced reviews, too. One was at mess lunch with other Air Force sergeants whose hands had been soiled with decades of jet plane grease and whose eyes had seen everything under the aeronautical sun. They suggested loudly that if the Eclipse project put a towload at the end of the C-141 where no designer intended one, the tail might break off. The polite phrases of the earlier memos in the offices conveyed the same message. But the mess-hall concern was more bluntly put.

The big reviews, however, were the formal ones. There was the PDR, a preliminary design review early in the project, followed by the CDR, a critical design review once 90 percent of the drawings had been done. As flight test drew near, there appeared the stern procedures of flight readiness review, the FRR. If the FRR was hurdled, its panel members, not the project members, presented it to the Airworthiness Flight Safety Review Board. If that was cleared, a project could fly. But in the case of Eclipse, there were other significant reviews. One was the video conference in the early summer of 1997 involving Dr. Robert E. Whitehead, the NASA Associate Administrator who headed the Office of Aeronautics and Space Transportation Technology. He gave a thumbs up to the project, satisfied the group knew what it was doing.

The least expected review came last. It was done by something dubbed the IRT,

[71] Robert Plested, interview by author, 11 August 1999.

[72] Dwain Deets, interview by author, 4 August 1999.

the Independent Review Team, an assessment group called into being by NASA Administrator Daniel S. Goldin. At the time, the event was the exception to the rule. In every review, the Eclipse team proved its case. Reukauf, Forger, Bowers, Collard, Murray, Lokos, and sometimes Gera devoted countless hours over these months to proving what they wanted to do was safe. Looking back, Reukauf thinks it was a good exercise, one that thoroughly rehearsed them all in the procedures to come.[73] And Forger, often impatient at the sheer number of presentations, in retrospect, also agrees.[74]

* * *

Late in August 1997, the Eclipse project gathered impetus. As the first day of actual flight research with the F-106 in the Eclipse configuration drew near, the most junior member of the team sat down to check some figures. He was Todd Peters, a structural loads engineer. He brought up on his computer screen the finite-element stress analysis of the F-106 fuselage and looked closer.

To an outsider, the image might have seemed lovely. You can see similar images in the opening montage sequences on *Discovery Channel* science shows where some real-world object is transformed into geometrical lines. The Eclipse project analysis displayed a vision of the F-106 fuselage reduced to geometrical patterns. The purpose of the finite-element stress analysis was to discover how much stress the F-106 fuselage could bear. The analysis had occurred long ago. KST had subcontracted the work in the days before NASA assumed test responsibility. The Dryden machine shop had already finished most of the work the analysis

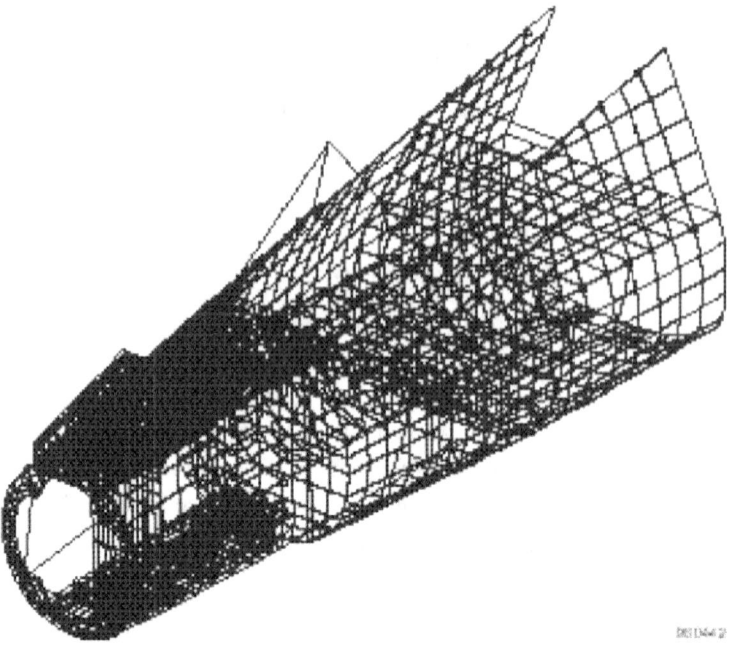

had indicated. But as Peters stared at the image on the screen, his eyes grew wide. The image of the finite-element stress analysis suggests precision and mathematical certainty, but the paradox remains that every line is also, in some sense, false. The model is really an illusion–a deft engineer can manage the trick, which is to combine these illusions into a sum that produces something true. For example, when the F-106 nose was analyzed, the sub-contracted engineer simplified its structure into a model to get his results.

"It's important when you simplify," explains Peters, "that you don't simplify an area that is crucial. If you do, the analysis can show everything is fine when it is not."[75]

According to Peters, the model seemed to have integrity on the screen when, in fact, it did not. And the next day, he took a signal step. He sent a young co-op engineer, Mike Allen, down to the hangar

Figure 3. Finite element model of the forward fuselage of the QF-106. (Design 980442 by the Dryden Graphics Office)

---

[73] Reukauf interview.

[74] Stucky interviews.

[75] Todd Peters, interview by author, 23 June 1999.

with tracing paper, white paint, flashlight, and calipers. Allen and a colleague would map the rivets and structural supports in the forward fuselage. As Peters continued his re-analysis at the computer, Allen sweated over real metal. There was a tiny hatch on the side of the airplane's nose. It was so small Allen could only reach his hand holding the caliper inside and awkwardly peer around his arm to see. From the hatch on the other side, another co-op shined a flashlight. By millimeters, they charted.

Allen, a polite, shy young engineering student, tells the story of all the rivets measured and today smiles and recalls, "I think everything was in good working order." But he also recalls late one afternoon during this intense period when Peters stopped by his desk.

"How are things going?" asked Allen.

"We're in deep, deep trouble," came the answer.

The crux was this: the F-106 was a lightweight airframe attached to one big engine. Not surprisingly, its nose was not designed for any tow load, let alone 24,000 pounds. According to Peters, what had gone wrong with the analysis was complex. It turned out that some modifications done to the airplane were not necessary, some were done incorrectly, and some important issues had not even been addressed.

There was a joint in the longerons (support members) in the nose. If you ran a finger, for instance, along the longeron, you would feel the break (a bolted joint that had been overlooked during visual inspection against airplane drawings), but this important reality did not appear in the finite-element stress analysis and thereby its author gave a forgiving nod to loads up to 24,000 pounds. But the joint would not support that substantial a load. According to Peters, that area of the fuselage might have failed at loads well below 10,000 pounds.

A complex web of complications resulted from such mistakes. If Peters were right, a possible scenario turned out to be the one several veteran pilots had fretted about early on–a mishap at takeoff when airplane and pilot were most vulnerable, some incident angle where the stresses on the fuselage later in fact did peak at 18,000 pounds, a catastrophe when just as the F-106 lifted into the air, its nose broke off.

Peters reported his findings up the chain of command. His superiors were not happy. They were all ready to fly. Suddenly he became an Issue–or felt he was one–at a time when everyone on the project wanted to be a non-issue and get in the air.[76]

The next day the managers scrambled. A phone call was put through to Bob Keltner at KST, and there were yowls of disbelief and pain on that end. Keltner called the subcontractor about the analysis that was being questioned, but the subcontractor had no answers, for he had subcontracted the task to someone else who could not be reached. What to do? How to figure this? There was no answer to these untimely questions.

NASA assigned Mark Lord to join Peters in the task. Lord was more easy-going than Peters, a quiet engineer mellowed with a generation of experience. Lord began re-doing the analysis with a pencil. Engineers call this approach classical analysis. It did not replace the analysis Peters had done with the software NASTRAN. Rather, it looked at the fuselage from a different angle and in a sense focused more closely. Both analyses, of course, were deft illusions

---

[76] In an editorial comment, Reukauf makes the point that the project was grateful to Peters, although he may have felt he was an issue.

aimed at understanding something real. At first, Lord's work with the pencil seemed to contradict that of Peters. Yes, of course, the analysis by the subcontractor had no validity, but Lord felt that the results would turn out, in his polite term, "beneficent."[77] But as Lord probed further, he too encountered serious problems.

Lord and Peters worked together, moving back and forth between their analyses, comparing, putting in a grueling seven-day-a-week, 7-a.m.-to-11-p.m. push to get the answers. The result was that the team did have to fix the fuselage. Rivets needed to be added to reinforce what had been incorrectly done. Lord designed metal straps to hold together questionable panels on the fuselage.

As the winter holidays of 1997 approached, the team raced to get finished before the Air Force put its tow airplane on the shelf.

Space

Space: defined in dictionaries as the region beyond the Earth's atmosphere. Of course, exactly where atmospheric particles thin out to virtual nothingness is subject to interpretation. But NASA had put a number on it, defining space by the international standard as a region beginning 62 miles above the surface of the Earth. It was a yardstick that decided who was an astronaut. The Air Force, on the other hand, had chosen to define space as a region 50 miles off-planet, awarding astronaut wings to X-15 pilots who ventured that high but not up to 62 miles.[78]

Space: its definition was not crucial to intellectual property in Kelly's patent. But it was the goal. And that December even as the Eclipse team raced to fly its tests over the desert, some KST engineers were asking themselves about modifications to make to the F-106 afterwards that might take it higher. Perhaps space wasn't so far away.

Kelly shared this enthusiasm. He had memories of watching Apollo flights on television as a child. While still an adolescent, he had penned an unpublished novel based on somewhat-real-world technology about teenagers flying to the Moon. Yet although Kelly was a visionary, he was also a very practical engineer. Hadn't retired Air Force Lt. Col. Jess Sponable, himself a hard-bitten realist in aerospace, suggested that all that was needed for economically feasible space flight was a reconfiguration of what had already been invented? "What America needs is not newer launch technology," said Sponable, "but today's technology applied to RLVs designed to fly with aircraft-like efficiencies."[79]

Al Bowers shared and shares his dream of space travel. His very office is something near a museum stacked with mementos of aerospace history, of the human race's effort to escape the gravitational pull of Earth. As a child, Bowers had watched with excitement the Apollo missions on live television. Despite his heavy workload at the center, he continued to donate time to public schools, talking about space exploration. But when he mentions the Apollo missions in his presentations, most of the school children have no idea a human being ever stepped on the Moon. It did not happen in their time.

---

[77] Mark Lord, interview by author, 15 July 1999.

[78] Dennis R. Jenkins, *Hypersonics Before the Shuttle: A Concise History of the X-15 Research Airplane* (Washington, DC: NASA SP-2000-4518, 2000), p. 61.

[79] Kelly interview; Lt. Col. Jess Sponable (USAF, Ret.), "The Next Century of Flight," *Aviation Week & Space Technology* (24 May 1999): 94.

And as NASA budgets dwindled from their levels in the Apollo era, as time passed and the only humans on the planet to visit the Moon became gray-haired members of the retirement generation, Bowers had a sharp sense of the rope's importance. Behind the tiny Eclipse Project wavered a question. When Neil Armstrong set foot on the Moon's surface in July 1969, whose shoes did he walk in?

Was it Leif Ericsson's?

Or Christopher Columbus's?

The Proof

On a bitter, cold Saturday morning in December, Forger ran his eyes over the gauges in his cockpit. This was the day. They were ready for flight test. His pilot flight cards clicked against an aluminum cockpit panel. The vast web of possibility had been refined to these simply printed cards. Here he was parked behind a C-141A, the stench of its fumes biting his nostrils. As he later reported, there was something unsettling in it all–despite his experience flying formation and flying refueling–something that seemed a violation of the most elementary commandment: never get behind a big transport on take-off.

Pilots snapped the flight cards on a kneeboard mounted on the left thigh. The cards were stiff, laminated, about the size of wine lists at restaurant tables. Typically, they had four punches in the left margin, the holes sometimes obliterating parts of words. They had indexes displayed along the bottom.

Forger knew many of the passages by heart. He knew the test sequences to come, the engineer commands such as "Cleared for pitch doublet!" that would be transmitted from the control room. He knew the emergency procedures, the most dire directives on take-off, the five steps of "abort" leading to the sixth: "Follow FLAMEOUT LANDING PROCEDURES."[80]

Delays had stalled them. It was a Saturday morning, 20 December 1997. Three weeks down-time lay ahead of them, two weeks for the holidays and a third week that annually shut down all projects for safety workshops. Could the Eclipse project squeeze in one flight test before the long layup? The Air Force's "drop-dead" date for the C-141A–February 1998–would not be extended. Unfortunately, Dryden Maintenance had decided that although the center director might give them special dispensation for a test the Saturday before Christmas, it was not likely to happen. On this assumption, the technicians had not fueled the safety chase airplanes ahead of time. Merry Christmas, Eclipse! The crew waited 30 minutes in the cold for refueling.

Finally, with all airplanes fueled and in position, the rope truck had done its work laying out the line, a carefully planned procedure carried out by a world-class crew that had trained itself for hooking up the rope without any abrasive damage. It was a cautious thousand-foot march down the runway between the two airplanes. Daryl Townsend, the big, easy-going crew chief was in front, followed by a technician with what looked like a shepherd's crook that he deftly maneuvered to keep the line from snarling on the nose of the F-106 and slapping on the concrete. The Air Force comedians liked to call this exercise "the parade of the Pharaohs."[81]

---

[80] Eclipse Project test cards, unpublished (see an example, document 21).

[81] By this, apparently they meant that the cautious walk of the NASA ground crew down the runway with the technician in back maneuvering what NASA folk described as a "shepherd's crook" wielded to keep the rope from slapping down, resembled the stately marches depicted in Egyptian art where some god or some ruler walked holding aloft a rod which often had a curved neck.

The minutes to come were crucial for Forger. He had to avoid a rope slack that might somehow initiate the worst-case scenario, two airplanes trying to take off with the connecting line serpenting around their gear. He also had to avoid too sudden a tug that might damage the towrope or lead to a break in the frangible link and shut them down. The week before on the taxi-tow test (an exercise limited to runway work), he had a problem. He was too quick on the brakes as the Starlifter began its slow acceleration. The F-106 jumped forward, the rope slapping on the ground, and in a blink he counted three oscillations before he controlled it, just hoping the control room would not call abort. He did not intend to let that happen again.

At a distance of a thousand feet, the C-141A was in position, gleaming from the early sun, and to his right and left, Forger saw the rope technicians, clear now from the flight path, their breaths pluming in the air, shifting from one foot to the other, still poised even when nothing was left for them to do, as if something still might be needed.

Forger's knees braced. If all other Eclipse procedures were carefully rehearsed science, these moments tensioning the rope with his feet on the brakes were close to art.

The large transport moved. The big jets shuddered and roared, and he watched as the rope slack started taking up. He worked his brakes. The important thing was not to slap the rope, not to start some oscillation. The rope seemed to draw almost gently off the tarmac and then cleared at 5,000 pounds just as Bowers had predicted from the very first. He called for Farmer to hold position at 6,000 pounds of tension, the rope straight as a ruler edge.

"Arris[82] ready for flight," rasped Farmer's voice on the radio.

"Eclipse ready for flight," answered Forger.

"Roger 20 seconds," crackled from the Starlifter. For the next 20 seconds, everyone on radio in the control room, as well as Forger, would sit in silence. He could hear the whine of the chase jets circling past, ready to move in. Then he could hear Drucker's voice, curiously modulated by radio from the tail of the C-141A, counting down, "Eclipse . . . 5 . . . 4. . . ."

When LeVake called "Brake release," the transport really started moving. The jets roared louder. "Smooth the brakes, smooth them," Forger thought as he relaxed his feet.

The "Parade of the Pharaohs." (NASA photo EC98 44393-32 by Carla Thomas)

---

82 The C-141A's radio call sign.

Eclipse project QF-106 and C-141A take off on first tethered flight 20 December 1997. (NASA photo EC97 44357-8 by Tom Tschida)

He was moving.

Suddenly, the tarmac blurred. He was racing down the oil-stained history of Runway 04. The yellow taxi marks whizzed past. The misty chain of mountains separating desert and sky waited in the distance. Thousand-foot markers, tin sheds fled past him. His gauge said 120 knots. When he hit 140 knots, he rotated the airplane to 7 degrees nose-up for

takeoff. But as he continued gathering speed, he realized there was no radio. Where was the Air Force? He could see the big bird above him lifting, climbing out as steeply as it could. He could feel the wake turbulence. But where were the radio calls rehearsed as the C-141 passed 100, 200, 300 feet to cue him for takeoff? This omission was not a "Red Light," not a required abort. But his pilot card advised he might choose to abort. "Follow the rope" had been the advice from KST and other joshing veterans at the base. And he did.[83]

The lift-off came with the Starlifter quite high in the envelope of operations. Farmer later commented he felt on this flight as if he dragged the interceptor off the ground. But Forger was off the ground.

"Eclipse airborne," called Forger.

He heard a familiar squawk. The Air Force came back on the radio. Down

Eclipse project QF-106 and C-141A climb out under tow on first tethered flight, 20 December 1997. (NASA photo EC97 44357-13 by Tom Tschida)

---

[83] Stucky interviews; report of Chief Engineer Al Bowers on Eclipse Flight 5 (1st towed flight), 20 December 1998 in Eclipse Flight Report (see document 6).

below cheers erupted in the packed control room as if a team had scored in a sports event. Forger realized he flew near the bottom of the planned low-tow area, and he climbed a few degrees. The good news was this: as he rolled out behind the C-141A and circled the eastern shore of the dry lakebed, he tracked very nicely, almost without pilot input. He continued the tests, step by step edging the F-106 to different areas beneath the tow airplane. Control was excellent.

This first tethered flight was a triumph. And at 10,000 feet as Forger was pulled by the Starlifter into a 40-degree roll, Mark Collard and Al Bowers stared in amusement at the video monitor in the control room, its image transmitted from the chase airplane. The engineers saw Forger seem to rise up from his seat.[84]

"He's not going to do what I think he's going to do, is he?" asked Bowers.[85]

But he was. He really was.

Forger raised both hands free of the aircraft's controls. The F-106 flew a smooth course.

"Forger," advised Collard over the radio, "if you are going to do this, move your hands so the camera can see." And the pilot clenched his fingers and waved his fists. To anyone who had labored through all the doubts, the briefings, the reviews, it was clear this moment was not show-boating. It was validation.[86]

Mike Kelly remembers, too, and recounts the story now without any note of I-told-

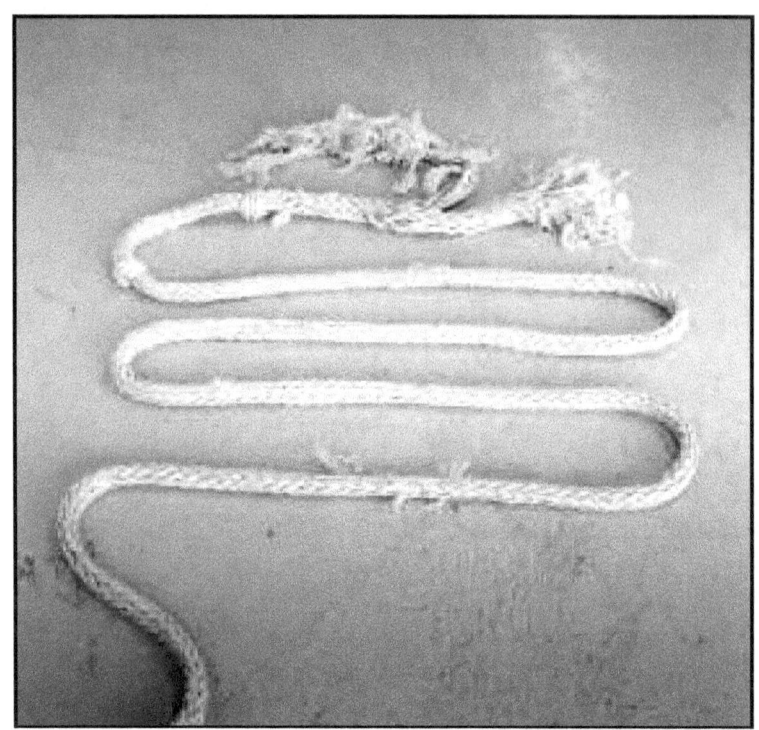

you-so. He sums up the first flight tests. "The only surprise was that there were no surprises," he says.[87]

* * *

But Kelly was forgetting about one incident that no one had foreseen. As Forger was flying behind the C-141 on the first flight, abruptly the rope released at his end. The whole 225 pounds of Vectran® and metal for a few moments became a violence in the sky. The partici-pants were unprepared for what they saw–a vast flailing–and on the second whipping, the metal knuckle snapped free and rocketed off into the blue.

Recovery of the knuckle remained a great hope for several days, and it became an extracurricular project. The pilot and

Tow rope after being whipped around by the knuckle assembly on the first tethered flight. (NASA photo EC97 44357-23 by Carla Thomas)

---

[84] As Stucky commented on the first draft of this monograph, "I didn't really stand up. I was, after all, strapped into an ejection seat. I simply twisted in my seat towards the chase video aircraft, raising my hands up over my head, and waved them when they asked to see some motion."

[85] Albion Bowers, interview by author, 25 June 1999.

[86] Interview with Collard.

[87] Kelly interview; cf. the documents on Eclipse Flight 5 in Eclipse Flight Report (see documents 3-7).

engineers launched a search mission. Using GPS data from the test flight, Forger flew over the area in a small airplane while below him rumbled several of the Eclipse people in their off-road vehicles, tracing grid-patterns across the desert wastelands. The task proved difficult. "If you look out there, it seems nice and flat, but when you drive in, it's little ravines and creek washes," said crewman Randy Button, who joined the pursuit.[88] "It's a big desert," smiled Kelly Latimer, nodding with irony.[89] But optimism prevailed for a time. Even a week later, a team member made one last Saturday sortie driving a jeep, wielding a GPS and metal-detector. They never found the knuckle, however.

The knuckle had been, of course, a flying weapon. According to Bowers who has a genius for predictive numbers, the knuckle might have been sailing at 300 miles per hour and possibly have buried itself in the sand ten feet deep. When he filed the pilot's report, Forger made 19 recommendations. Number 12 read: "Recommend future operations immediately occur in the Precision Impact Range Area (PIRA) airspace over uninhabited areas."[90]

Post-flight inspection revealed nothing wrong with the F-106 hardware, and post-flight data analysis showed no sudden stress on the tow line. The Dryden engineers could tell you of past test flights where something went awry. Reports would be filed that some*thing* did not "function." But sometimes it has been the pilot who has not functioned as intended.

In an afternoon meeting, Forger spoke up. He offered the opinion that he had inadvertently released the rope. The engineers had situated the pneumatic release button on the pilot's control stick. When Forger had his hand on the stick, his index finger rested a hairline from this button. He must have touched the button. His honesty here became part of the Eclipse story. If he had remained silent, everyone on the project half-guessing the release scenario, it would in some way have fed the worry that it could happen again. Immediately the engineers offered to move the button elsewhere–perhaps its placement had not been a great idea. Forger said they did not need to lose the time on installation. Daryl Townsend remembered how the pilot's eyes grew narrow. "It won't happen again," he said.[91]

And it did not.

When the Starlifter towed the F-106, the Air Force pilots could barely tell they had a tow load. It was a subtle difference. In flight, they could not see the F-106, nor did they have video display. When Forger released the rope, the Air Force pilots felt a gentle surging forward, nothing else. On the second test flight, Farmer claimed to hear a low noise that he thought was transmitted through the rope, a low-frequency rumbling that disappeared after the rope release. "Sure," grinned co-pilot Latimer, intoning her disbelief.[92]

---

[88] Randy Button, interview by author, 25 June 1999.

[89] Kelly Latimer, interview by author, 6 July 1999.

[90] Pilot's Flight Test Report, First Tethered Flight in Eclipse Flight Report (see document 4).

[91] Townsend interview.

[92] In editorial comments, Stucky wrote, "The quivering/shimmering the canvas sleeve caused to the towrope was, I feel, the source of the rumbling in the rope that was transmitted to the C-141 and [that] Farmer could feel." See below for discussion of the canvas sleeve.

But Latimer and Farmer felt some twinge of regret hearing the Air Force crew in back at the tow connection, shouting in astonishment, "You should see this! This is the coolest!" It was the rope they saw. It was moving in wild, beautiful oscillations nobody had predicted. Sergeant Dana Brink, the scanner, recalls, "Once the rope disconnected, it was like when you take and whip a garden hose, at first the curve gets bigger and slower at the same time." Brink and Drucker both viewed the writhings at close hand, and at a flight debriefing when Eclipse engineers said wistfully they wished they had a better look at the rope (its thin stripe difficult to see by eye or chase video against the glaring desert sky), Brink admitted he had unauthorized photography.[93] According to several project members, the sergeant's pictures were the most stunning images recorded during the experiments.

* * *

In January, when the Eclipse team had returned to flight research, the drop-dead date still loomed ahead for the C-141A, and for practical purposes, the group could not work past 6 February 1998. On 21 January, the airplanes ascended for their second tethered flight.

The engineers had anticipated a flight envelope of easy operation, one that proved easier in real flight than the sims predicted, so easy that the rope, which in fact bowed, might as well have been straight in certain low-tow configurations. But Forger had parameters to explore in these tests. There were places Forger tried to fly where the F-106 "turned into a bucking bronco," and he brought it back. At another point in high tow, he spiked his control stick (an abrupt control input to see if the airplane would return to

View of the F-106 and the tow rope from the C-141A. (NASA photo EC98 44393-52 by Carla Thomas)

---

[93] Dana Brink, interview by author, 1 July 1999.

Side view of the C-141A towing the F-106. (NASA photo EC98 44415-19 by Jim Ross)

stability or oscillate). Suddenly, everything changed. The idyllic curve of the rope was turned "to an unnerving and chaotic spaghetti-like appearance."[94] It was, he said, a place he did not want to be, but it was part of the test. He immediately pushed over and descended to the safety of the low-tow position.

During ground tests, the team had anticipated that 50 feet of nylon attachment at the C-141A end of the rope might create a problem. The engineers had moved this segment to the middle of the Vectran® tow line to damp the oscillations. They had covered this 50-foot damping section with a canvas shroud to protect the nylon during

hookup on the runway. The canvas was too big, and on the first flight, it began tearing apart in the air.

"At first," recalls Forger, "I could see it just quivering." Then objects flung past him. "I could see what came off," he says, "I could see it fly."[95]

On the second flight, the crew taped the canvas down tightly, but during the experiment it was "shimmering" and interfering as an aerodynamic factor in the tests. Finally, from the third flight on, the engineers decided to do away with the nylon in the middle–Vectran® could handle the damping.[96]

---

[94] Mark Stucky, interview by author, 22 July 1999.

[95] Ibid.

[96] See Pilot's Flight Test Reports, EXD-01 Flight 5—First Tethered Flight; Flight 6—Second Tethered Flight, 21 January 1998; Flight 7—Third Tethered Flight, 23 January 1998, all from Eclipse Flight Report (documents 4, 8, and 15).

As Murray had guessed, there was a dynamic in the tow line. Whenever Forger made large control inputs to the F-106, the Air Force pilots could feel the effects through the rope. But Murray wanted to know more about the loads on the rope. All of the measurements were being made from Forger's end of the rope. The technicians now raced for a new measurement procedure involving a load cell signal taken at the C-141A end, which would be recorded on a modified laptop computer and monitored by a technician seated in the rear of the transport. They began gathering this data on the fifth flight.[97]

No one had yet put a mathematical model on the real ferocity of the rope.

* * *

Strong Pacific storms came blowing toward the desert on Thursday. The last Eclipse test had been approved for Friday, 6 February. It was a crucial test–to date, all the flights had ascended no higher than 10,000 feet, but if you listened to engineering anecdote, interesting things could start happening on tow at 25,000 feet. Kelly's concept included towing at these altitudes. The team needed to do this experiment.

Winter is always the worst season at Edwards, the pilots say. And this was the El Niño year. When the engineers, managers, and pilots looked up at the TV

weather report that Thursday, they saw graphics of a vast cloud cover arriving. The Air Force forecast was rain starting at dawn on Friday, low visibility, and winds gusting to 30 knots. Test flights simply were not done in these conditions. The Eclipse team was limited to winds of less than 15 knots for takeoff, 10 knots for a tailwind.[98]

Casey Donohue, the young Dryden meteorologist, recalls, "The El Niño front actually was going northwest to southeast." He explains that it seemed to him the lower end of the storm crossing California would tend to snag on the mountains.[99]

Tow rope after second tethered flight. (NASA photo EC98 44390-24 by Carla Thomas)

---

[97] The numbering of the flights is somewhat confusing because Forger had flown the F-106 alone (without the C-141) in the Eclipse (EXD-01) configuration four times in October and early November 1997 to calibrate air data and validate simulations (see document 1). Then the taxi test on 13 December counted as an Eclipse mission, making the first tethered flight on 20 December the 5th flight in the EXD-01 configuration. Thus the 5th tethered flight mentioned above in the narrative was actually the 9th EXD-01 flight. It took place on 5 February 1998, with the last flight (#6) the following day. On the load cell, see Project Manager's Comments, Fifth Tethered/Release Flight, 5 February 1998 (document 30), Bill Lokos' Structures Report (document 36), and Jim Murray's Flight Mechanics Report for the same flight (see document 35), all in Eclipse Flight Report. Dryden's Mark Nunelee prepared the load cell and Allen Parker expeditiously set up the laptop computer.

[98] Taking off with a tailwind is generally unacceptable. However, the Eclipse team established a tailwind limit because it was safer to take off to the East (toward the lakebed for possible emergency landings) and prevailing winds at Edwards AFB are from the West. Comments of Carol Reukauf on a draft of this study.

[99] Casey Donohue, interview by author, 12 July 1999. See also Casey's Weather Summary for Eclipse EXD-01 Flight 10, 6 February 1998 in Eclipse Flight Report (document 49).

The Eclipse team met Thursday at 3 p.m. It was grim. After Saturday, the Starlifter was gone. All of the weather forecast services had agreed on the Friday forecast.

At the meeting, Donohue took a deep breath. Then he said, "There's going to be a window."

After deliberation, the team decided not to scrap the flight; they would get in their cars and drive out very early tomorrow; they would see. But was Donohue right? Forger had a personal contact with the Air Force weather service. He decided to get some up-to-the-second forecasting. He made the phone call.[100]

Bowers remembers the event–while he and Forger waited for the answer, over the phone they could hear the Air Force forecasters laughing. The answer came, but Forger hesitantly interrupted.

"We hear," he said, "there's going to be a little break tomorrow morning."

"Yeah, there's going to be a break out near Hawaii."

That night, Bowers did not sleep well. He typically does not when the excitement of the test looms, when he knows he will rise early. The clouds had already appeared the evening before. He recalled, "I got up at 2:30 next morning and went outside and looked up and there were stars and I knew we were gonna fly."[101]

In a world of metal and instrumentation, the Dryden people seem to take precision for granted, but weather is something else. When Bowers told about Donohue's prediction, he leaned forward in his chair. "He hit it exactly," says Bowers, gleaming in admiration.

The next morning the Eclipse aircraft took off on Runway 22. It was not perfect weather. There was broken cloud cover. The self-effacing Donohue will try to tell you that his forecast of winds switching to the south proved incorrect; hence, they should have used another runway, as they had on all previous flights. The final reports list winds at 11 and 12 knots, higher than the tailwind limit of 10 knots. But because they were not direct tailwinds, they were within limits. Consequently, the C-141A pulled the F-106 up into the air.

It was an adventure. Typically test flights are not supposed to occur when visibility is under three miles. "You want to see the ground, everything," explained an Air Force aviator. That Friday, the possibility of maintaining VMC (visual meteorological conditions) did not look good. Gordon Fullerton, called Gordo by many of his associates, was chase pilot. Early on in the project, he had been skeptical about safety issues, but now he volunteered "to go up and take a look around." He came back with the message that he thought the flight should be a go.

The airplanes took off and began to climb. Farmer remembers the view from the cockpit of the Starlifter. He looked up and saw what seemed nearly unbroken dark clouds. He followed Fullerton, who was leading the way in the chase airplane to holes in the cloud cover.

"You know this sounds corny," Farmer explained–seeming embarrassed as if he might hear from the Air Force joshers for these remarks. But he did not stop. "I've never seen such a thing before or since. It was like magic that day, the way holes opened up in the clouds and Gordo flew through, and I followed tugging the F-106 along

[100] Stucky interviews.

[101] Albion Bowers, interview by author, 17 July 1999.

behind. We were flying through the sky looking for holes in the clouds."[102]

They ascended to 25,000 feet. Forger performed his maneuvers. The rope contributed some instabilities, but the F-106 was flyable. He elected to remain on tow for part of the descent, to gather more data, which he did with speedbrakes deployed so that their drag would maintain rope tension and keep the airplane stable. He had been releasing recently from tow by breaking the frangible link. This time at 9,000 feet, he simply released the rope. They would not need the knuckle again. The long line snaked, and on its first whip, cast the knuckle into the desert. When Forger landed, he was still taxiing on the runway as the first rain began to pepper the concrete. Strong gusts began tossing wind socks. The El Niño storm had arrived.[103]

Mike Kelly watched from the flight control room. "This was quite a triumph," he said. "Here was this project everyone thought was unsafe in the beginning–no one should fly this–and now despite adverse weather conditions, people had all this confidence in towing technology."[104] Cheers broke out across the control room. People clapped and applauded for the many heroes on the team. And Bob Keltner threaded his way across the noisy room to shake the hand of the meteorologist.

* * *

Afterwards, the emphatic triumph of the tow demonstrations made aviation news.[105] The stills and videos taken from the NASA Dryden chase airplanes brilliantly documented what had been achieved.

The Eclipse project won the Team Project of the Year Award for 1998 at NASA Dryden. NASA Administrator Dan Goldin sent a note of personal thanks to Dryden Center Director Ken Szalai.[106] Many of the members of the team will tell you in retrospect that Eclipse was the most rewarding and exciting project they have ever worked on. In their offices, their scraps of rope hung as trophies proudly display this sentiment. "It came, we did it, it went away," said Jim Murray with a real sense of accomplishment.[107] "I came away with the memory," said one Air Force team member, "that if you keep plugging ahead, everything will work out." Several other team members echoed this sense of significant lessons learned. As this history was written, Jim Murray had been assigned to apply his brilliance to designing an airplane intended to fly on Mars in 2003. Al Bowers moved on to become chief engineer on the revolutionary Blended Wing Body Project.

With this success behind them, Mike Kelly and his staff moved on with their agenda. They had unfinished business. The Eclipse was merely one step on the path. For awhile, KST tried to talk the Air Force into letting it have several F-106s. Kelly had an idea for installing a rocket on the F-106. It would be a sub-scale

---

[102] Farmer interview.

[103] Pilot's Flight Test Report, EXD-01 Flight 10—6ᵗʰ (Final) Tethered Flight, 6 February 1998, in Eclipse Flight Report (see document 43).

[104] Kelly interview.

[105] See, e.g., Bruce A. Smith, "Tow Concept Tested," *Aviation Week & Space Technology* (9 February 1998): 93.

[106] Note, Dan Goldin to Ken Szalai, 2 April 1998 (see document 54).

[107] Murray interview.

aircraft. He stared back over his shoulder at the image and then did not say a word but climbed in the jeep and drove away.[109]

This unexpected meeting of the C-141A with another F-106 perhaps symbolized the unknown outcome of the Eclipse project. Would it lead to a new way to launch spacecraft? As these lines were being written, the answer was not

The F-106 taking off for its flight to Davis-Monthan Air Force Base, Arizona, after the Eclipse project ended. (NASA photo EC98 44534-02 by Tony Landis)

version of the dream, but they would actually use the airplane as a commercial-satellite launch vehicle. The Air Force refused. At the time this history was written, at its website KST was advertising for people to pay KST $90,000 now in order to be on the list for tourist launches KST had planned for the year 2001.[108]

One final and strange event happened with the C-141A. Pilot Stu Farmer tells the story. He had flown the Starlifter on its sad journey down to the air museum at Dover, Delaware. When the museum staff brought the C-141A to its final resting place, they parked it in front of an F-106. Farmer glanced a moment at the field attendant who seemed unaware of the significance (perhaps the irony) of the accidental juxtaposition of the two

clear. The project left its participants with a sense of accomplishment. The data generated might find multiple applications in the world of aeronautics and space. But the puzzles and urgency remained.[110]

In the end, the issue with the tests was neither the F-106—its aerodynamics were known—nor the C-141A, whose aerodynamics were also known. In the end, the rope was the crux of the matter.

Mike Kelly was clear about that. He shared what flew through Forger's head in the decisive moment on take-off, the advice from the old-timers on the base, an admonition, a remark that was both jest and truth.

Follow the rope.

---

[108] The URL for the website was http://www.kellyspace.com/

[109] Farmer interview.

[110] This paragraph is heavily indebted to J.D. Hunley, chief historian at NASA Dryden.

NASA Dryden Flight Research Center
History Office

Eclipse Project Flight Log

Compiled by Peter W. Merlin
June 1999

Preliminary Tests:

Flt. 01 / Test F1 / 24 OCT 96 : F-18 (161703 / NASA 850) was flown behind the C-141A (61-2775) to survey wake turbulence levels and determine the preferred tow locations for future F-106 tow operations.

Flt. 02 / Test F3 / 23 DEC 96 : C-141A (61-2775) was flown with wingtip smoke generators to study wake turbulence patterns.

Flt. 03 / Test F2 / 17 JUL 97 : QF-106A (60-0010) was flown behind C-141A (61-2775) to determine the preferred tow locations for future F-106 tow operations.

QF-106A/EXD-01 (60-0130):

Flt. 01 / 01 OCT 97 : Functional check flight, simulation validation, and rotational characteristics.

Flt. 02 / 08 OCT 97 : Airdata calibration and simulation validation.

Flt. 03 / 08 OCT 97 : Airdata calibration and simulation validation.

Flt. 04 / 04 NOV 97 : Airdata calibration and simulation validation.

Taxi 01 / 13 DEC 97 : High speed tethered taxi test. Release during EXD 01 rotation, and obtain C-141 takeoff performance data.

Flt. 05 / Test T1 / 20 DEC 97 : First tethered flight.

Flt. 06 / Test T2 / 21 JAN 98 : Second tethered flight.

Flt. 07 / Test T3 / 23 JAN 98 : Third tethered flight.

Flt. 08 / Test T4 / 28 JAN 98 : Fourth tethered flight.

Flt. 09 / Test T5 / 05 FEB 98 : Fifth tethered flight.

Flt. 10 / Test T6 / 06 FEB 98 : Sixth tethered flight.

QF-106A (60-0010) Flight Log

Compiled by Peter W. Merlin
October 1999

A modified production F-106A was obtained by NASA from the U.S. Air Force to support the Eclipse project. Mark Stucky ferried the aircraft to NASA DFRC from Mojave on 17 June 1997. Seven NASA pilots flew the aircraft over a 10 month period. Stucky ferried the aircraft to Davis-Monthan AFB, Arizona on 30 April 1998 for storage.

Pilots included:

Mark "Forger" Stucky   Dana Purifoy
James "Smoke" Smolka   Rogers Smith
Thomas C. McMurtry   C. Gordon Fullerton
Edward "Fast Eddie" Schneider

Flt. 01 / 17 JUN 97 : Stucky. Ferry flight from Mojave.

Flt. 02 / 15 JUL 97 : Stucky.

Flt. 03 / 17 JUL 97 : Stucky. Flown behind C-141A (61-2775) to determine the preferred tow locations for future F-106 tow operations.

Flt. 04 / 19 AUG 97 : Stucky.

Flt. 05 / 21 NOV 97 : Stucky.

Flt. 06 / 19 FEB 98 : Smolka. Pilot familiarization.

Flt. 07 / 03 MAR 98 : Stucky.

Flt. 08 / 04 MAR 98 : McMurtry. Pilot familiarization.

Flt. 09 / 17 MAR 98 : Schneider. Pilot familiarization.

Flt. 10 / 19 MAR 98 : Purifoy. Pilot familiarization.

Flt. 11 / 20 MAR 98 : Smith. Pilot familiarization.

Flt. 12 / 24 MAR 98 : Fullerton. Pilot familiarization.

Flt. 13 / 09 APR 98 : Stucky.

Flt. 14 / 30 APR 98 : Stucky. Ferry flight to Davis-Monthan AFB, Arizona for storage.

QF-106A / EXD-01 (60-0130) Flight Log

Compiled by Peter W. Merlin
June 1999

NASA obtained a QF-106A from the U.S. Air Force for use in support of Project Eclipse. Mark Stucky made 17 flights and one taxi test in the aircraft. It was then returned to the Air Force for storage.

Flt. P1 / 14 FEB 97 : Local area proficiency flight from Mojave.

Flt. P2 / 18 FEB 97 : Local area proficiency flight from Mojave.

Flt. P3 / 06 MAY 97 : Local area proficiency flight from Mojave.

Flt. P4 / 08 MAY 97 : Local area proficiency flight from Mojave.

Flt. P5 / 21 MAY 97 : Ferry flight from Mojave to NASA Dryden.

Flt. 01 / 01 OCT 97 : Functional check flight, simulation validation, and rotational characteristics.

Flt. 02 / 08 OCT 97 : Airdata calibration and simulation validation.

Flt. 03 / 08 OCT 97 : Airdata calibration and simulation validation.

Flt. 04 / 04 NOV 97 : Airdata calibration and simulation validation.

Taxi 1 / 13 DEC 97 : Release during EXD-01 rotation, and obtain C-141 takeoff performance data.

Flt. 05 / Test T1 / 20 DEC 97 : First tethered flight.

Flt. 00 / Test T2 / 21 JAN 98 : Second tethered flight.

Flt. 07 / Test T3 / 23 JAN 98 : Third tethered flight.

Flt. 08 / Test T4 / 28 JAN 98 : Fourth tethered flight.

Flt. 09 / Test T5 / 05 FEB 98 : Fifth tethered flight.

Flt. 10 / Test T6 / 06 FEB 98 : Sixth tethered flight.

Flt. 11 / 30 APR 98 : Functional check flight following restoration of aircraft to original condition.

Flt. 12 / 01 MAY 98 : Ferry flight to Davis-Monthan AFB, Arizona for storage.

United States Patent [19]

Kelly

[11] Patent Number:	5,626,310
[45] Date of Patent:	May 6, 1997

[54] **SPACE LAUNCH VEHICLES CONFIGURED AS GLIDERS AND TOWED TO LAUNCH ALTITUDE BY CONVENTIONAL AIRCRAFT**

[75] Inventor: Michael S. Kelly, Redlands, Calif.

[73] Assignee: Kelly Space & Technology, Inc., San Bernardino, Calif.

[21] Appl. No.: 342,596

[22] Filed: Nov. 21, 1994

[51] Int. Cl.^{6} B64D 5/00
[52] U.S. Cl. .... 244/2; 244/158 R
[58] Field of Search .... 244/158 R, 2, 244/3, 63, 118.1, 135 C, 135 R, 172, 129.5, 118.3, 161

[56] **References Cited**

U.S. PATENT DOCUMENTS

2,402,918	6/1946	Schultz	244/3
2,723,812	11/1955	Hohmann	244/3
2,823,880	2/1958	Bergeson	244/135 C
3,437,285	4/1969	Manfredi et al.	244/63
3,747,873	7/1973	Layer et al.	244/3
3,857,534	12/1974	Drees et al.	244/17.27
4,235,399	11/1980	Shorey	244/129.5
4,265,416	5/1981	Jackson et al.	244/63
4,646,994	3/1987	Petersen et al.	244/158 R
4,784,354	11/1988	Tavano	244/135 B
4,802,639	2/1989	Hardy et al.	244/158 R

FOREIGN PATENT DOCUMENTS

55-47658	12/1980	Japan	F16F 9/30
60-124384	8/1985	Japan	B62D 27/06
2-9988	3/1990	Japan	B62D 27/06
5-4472	2/1993	Japan	B62D 33/07

Primary Examiner—Andres Kashnikow
Assistant Examiner—Tien Dinh
Attorney, Agent, or Firm—Christie, Parker & Hale, LLP

[57] **ABSTRACT**

Orbital launch vehicles equipped with aerodynamic lifting surfaces enabling them to be towed as gliders behind conventional aircraft, and the method of towing these launch vehicles using a flexible cable to connect them with a conventional aircraft, for placing spacecraft into low earth orbit at greatly reduced cost compared to current orbital launch systems. The lift from the aerodynamic surfaces enables the launch vehicles to be towed by means of a flexible cable from a conventional runway using existing aircraft. As with "conventional air-launch," this permits spacecraft launch into orbit to originate from any conventional runway consistent with constraints of public safety, thus eliminating the need to build dedicated launch pads at geographic locations from which a full range of orbital inclinations can be reached. The method of towing the launch vehicle, utilizing the lift of its wings to fully offset its weight, permits at least an order of magnitude increase in the weight of vehicle which can be launched compared to "conventional air-launch" methods whereby the launch vehicle is carried on or within a conventional aircraft. This in turn enables an order of magnitude increase in the weight of spacecraft which can benefit from the inherent flexibility and low cost of "air-launch." The tow launch method also requires fewer and simpler modifications to a conventional aircraft than do any other current or proposed air-launch methods.

17 Claims, 4 Drawing Sheets

Document 2. U.S. Patent Number 5,626,310, assigned to Kelly Space & Technology, Inc., for Space Launch Vehicles Configured as Gliders and Towed to Launch Altitude by Conventional Aircraft

1

SPACE LAUNCH VEHICLES CONFIGURED AS GLIDERS AND TOWED TO LAUNCH ALTITUDE BY CONVENTIONAL AIRCRAFT

BACKGROUND OF THE INVENTION AND PRIOR ART

This invention relates generally to launch vehicles for placing spacecraft into orbit around the earth and, more particularly, to launch vehicles equipped with lift producing surfaces of sufficient capacity to permit the launch vehicles to be towed as gliders behind conventional aircraft. A launch vehicle so configured may be regarded as "air-launched" by conventional aircraft, or, alternatively as a launch vehicle augmented by a conventional aircraft which serves as a "zero-stage."

A limited number of differing types of launch vehicles is currently available for placing spacecraft into orbit around the earth. Virtually all are launched under rocket power from a fixed launch pad. This limits the rapidity with which launches can be performed to the time required to prepare the launch pad, assemble the launch vehicle on the pad, place the spacecraft on the vehicle, load propellant into the vehicle, verify that its systems are operating properly, and perform the launch. When the requirement arises to place a spacecraft into a specific orbital plane with respect to the fixed stars, the opportunity to launch is limited to a very short time as the orbital plane passes over the launch site. This time, referred to as the launch window, can be as short as a few seconds if the desired orbital plane is highly inclined to the equator and the launch pad is at a low latitude. If any operation leading up to launch is delayed, the launch window may be missed, and the launch may have to be delayed until the next opportunity. The complexity of launch operations is often such that the next passage of the desired orbital plane occurs before the vehicle can be made ready for another attempt. Maintaining a launch crew on site and repeatedly performing pre-launch operations is a significant contributor to the high cost of space launch operations.

Pad-launched vehicles can deliver spacecraft only to certain orbital inclinations by virtue of the geographic location of the launch pad. Safety concerns related to flying over inhabited land masses restrict the direction in which a vehicle can be launched from a given pad, and consequently limit the maximum inclination of the orbit which can be achieved. The minimum inclination which can be achieved from a fixed launch pad is determined by and equal to the geographic latitude at which the pad is situated. Though propulsive maneuvers can be performed to change orbital inclination once the spacecraft is in orbit, the weight of propellant required to do so is prohibitive for changes greater than 5 or so degrees.

Launch pad construction is very costly, as is launch pad maintenance and post-launch refurbishment. These costs are reflected in the cost of launch. The nature of the earth's geography is such that only a small number of remote locations, at the equator, are suitable for launching into orbits of arbitrary inclination. For launch service providers who do not have access to these locations, multiple launch sites at various locations must be built in order to be able to place spacecraft into orbits of arbitrary inclination. The cost of multiple launch sites can be prohibitive, so that launch service providers are unable to afford enough sites to launch into orbits of arbitrary inclination. This results in a restriction of the types of missions that can be performed by a given launch service provider.

A recently implemented improvement in space launch has emerged wherein the launch vehicle is carried on board a

2

conventional aircraft. The aircraft can fly to an arbitrary geographic location, where the launch vehicle is released and propels its payload (spacecraft) into orbit. This operation is referred to as "air-launch," and vehicles so configured as "air-launched."

An alternative way of regarding air-launch, appropriate when applied to launch vehicles capable of taking off from the ground, is to consider the launch aircraft as a "zero-stage." This parlance is commonly used to describe propulsion systems added to existing launch vehicles to augment their performance by raising them to a certain altitude and velocity before the launch vehicle's own propulsion system can be ignited. This reduces the total energy the existing launch vehicle must add to the payload, and translates into either greater payload capacity or into placing the same payload into a more energetic orbit. Reference to the launch aircraft as a "zero-stage" would apply in cases where the launch vehicle is either capable of taking off from the ground under its own power, or where the launch vehicle was not specifically designed to be air-launched.

The advantages of air-launch over ground-launch are numerous. The launch location can be selected so that no inhabited land mass is jeopardized by the vehicle as it flies over, yet the spacecraft can be placed into an orbit of any desired inclination. The variety of missions which can be performed using this aircraft as a launch platform is thus significantly greater than that which can be performed by a vehicle launched from a fixed pad. Moreover, only one aircraft need be purchased, and it can be flown from any conventional airport facility which will permit such operation. This is equivalent to having one "launch pad" (the aircraft) which can be easily moved to any desired geographic location. In the alternative representation of such a system as a launch vehicle having an aircraft as a zero-stage, the equivalence becomes one of having multiple launch pads already in place around the world in the form of the above mentioned conventional airport facilities.

Also, when launching into specific, highly inclined orbits, the aircraft launched vehicle can have a launch window whose duration is limited only by the time the aircraft can remain aloft. This can be accomplished by flying westward at a latitude and speed which permit the aircraft to keep pace with the orbital plane as the earth rotates beneath it. The chances of missing a launch window are thereby significantly reduced.

As mentioned previously, the launch vehicle has to add less potential energy to the spacecraft, since it begins its powered flight at a higher altitude than does a vehicle launched from a ground-based pad. The velocity of the aircraft is also added to that of the launch vehicle, so that the launch vehicle does not have to provide all of the velocity needed to reach orbit. If the launch vehicle is rocket propelled, the performance of the rocket engine can be higher than if it is launched from the ground due to the lower back-pressure on the nozzle at the launch altitude.

Finally, for a given orbital inclination, the launch vehicle may be launched in a due-east direction from a latitude equal to the desired orbital inclination. This adds the velocity of the earth's rotation to the vehicle's initial velocity to the maximum extent possible. These factors all contribute to a vehicle which, for a given launch weight, can place a heavier spacecraft into orbit than it could if launched from the ground, or the same payload into more energetic trajectories.

Even more performance enhancement is gained by adding lifting surfaces to the vehicle. These use aerodynamic forces to augment the thrust produced by the launch vehicle's

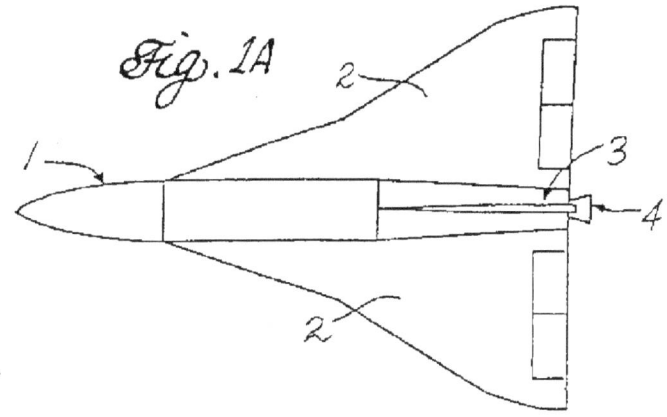

Fig. 1A

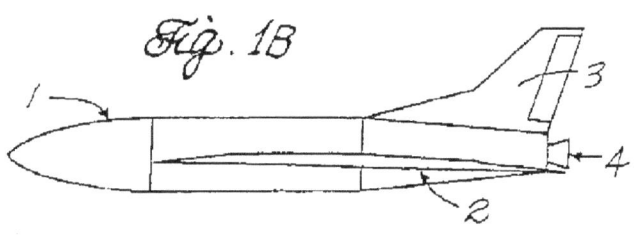

Fig. 1B

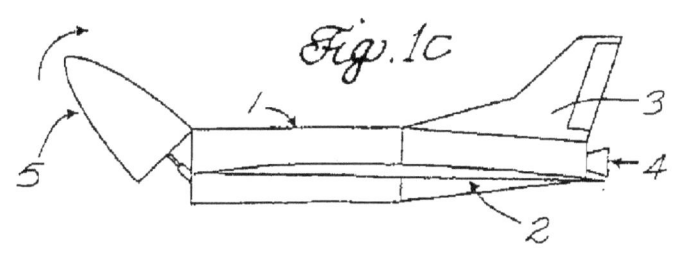

Fig. 1C

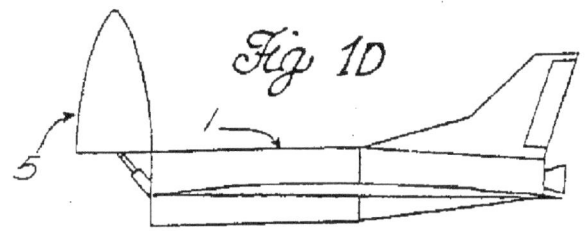

Fig. 1D

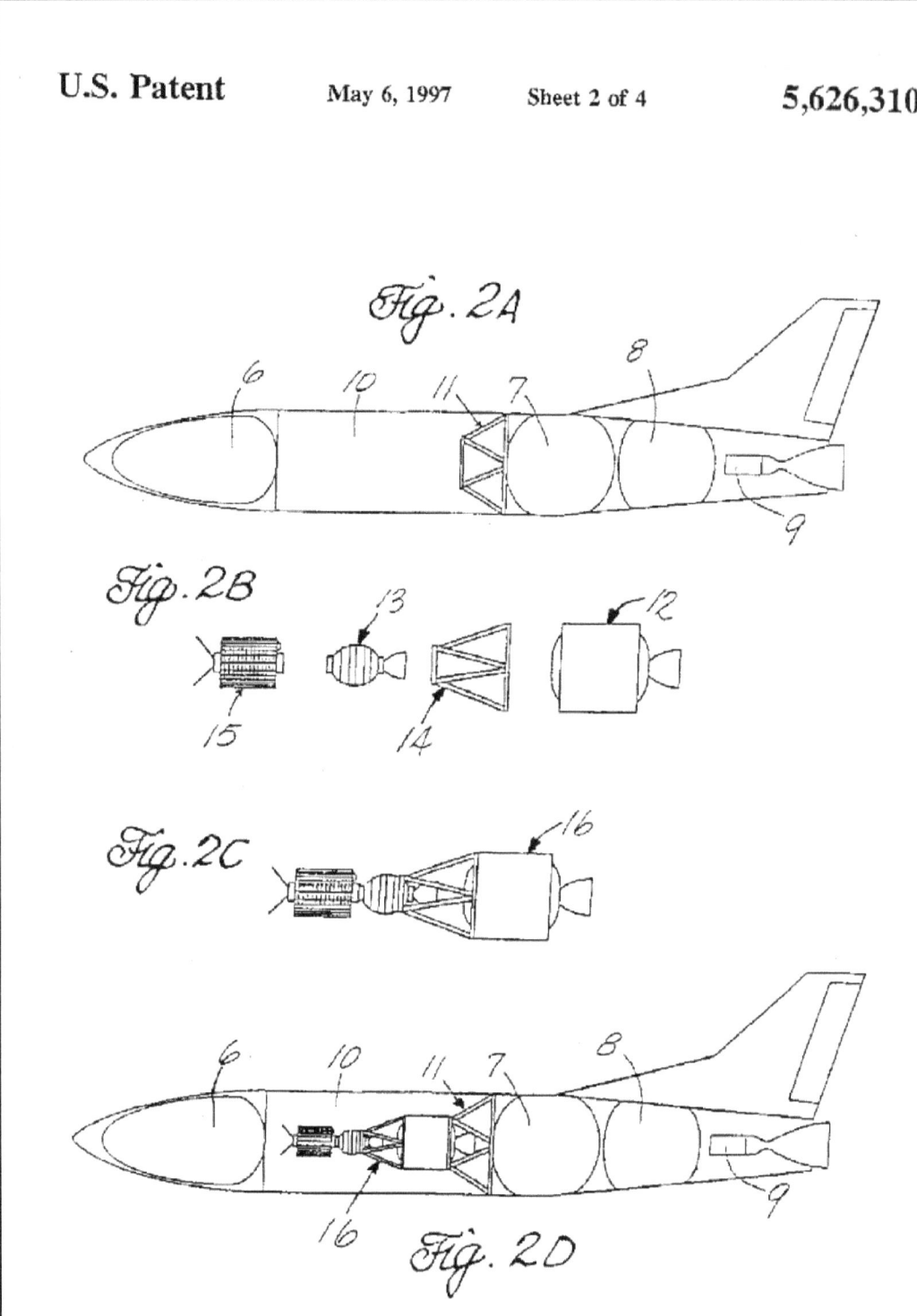

Fig. 2A

Fig. 2B

Fig. 2C

Fig. 2D

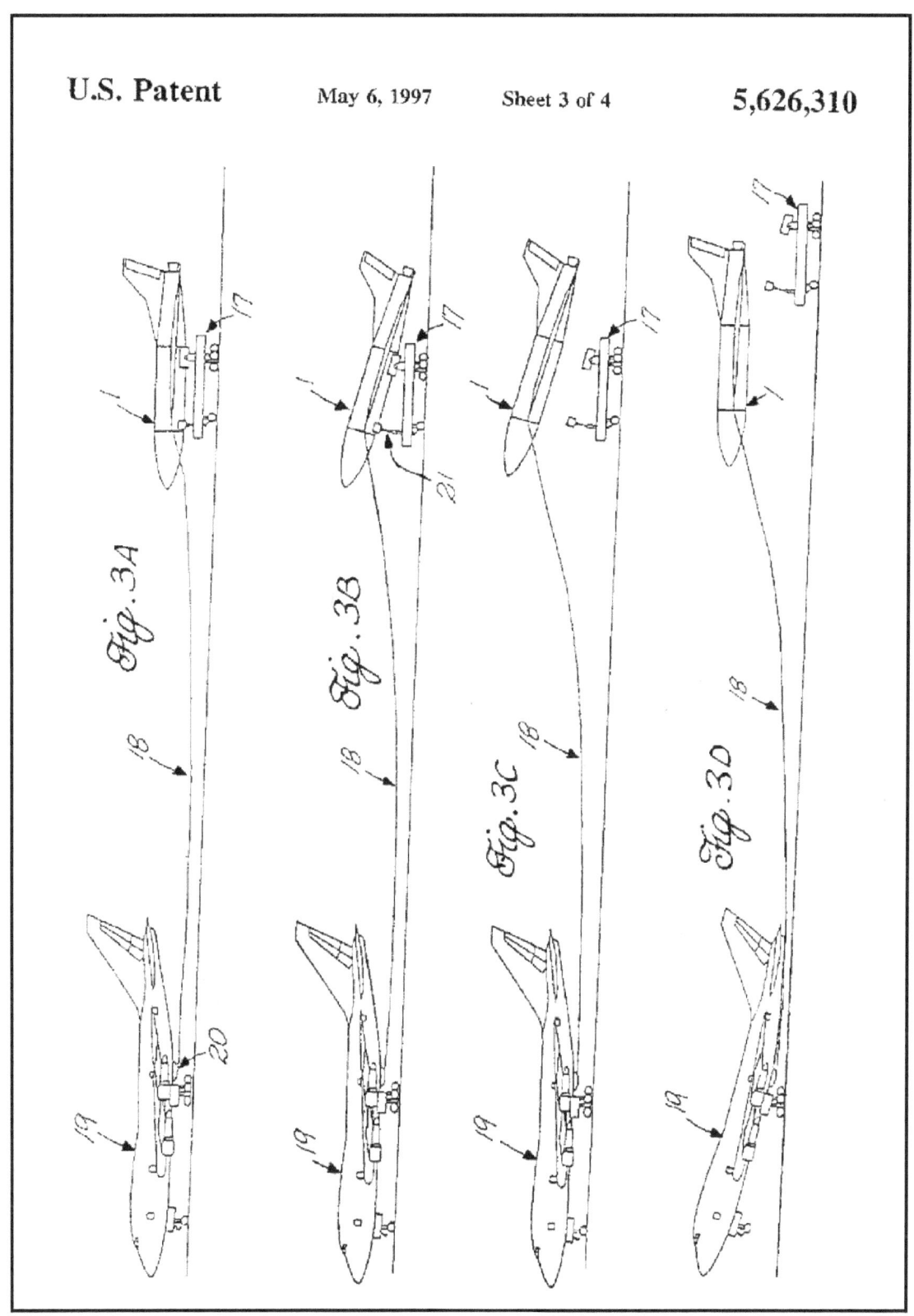

Fig. 3A

Fig. 3B

Fig. 3C

Fig. 3D

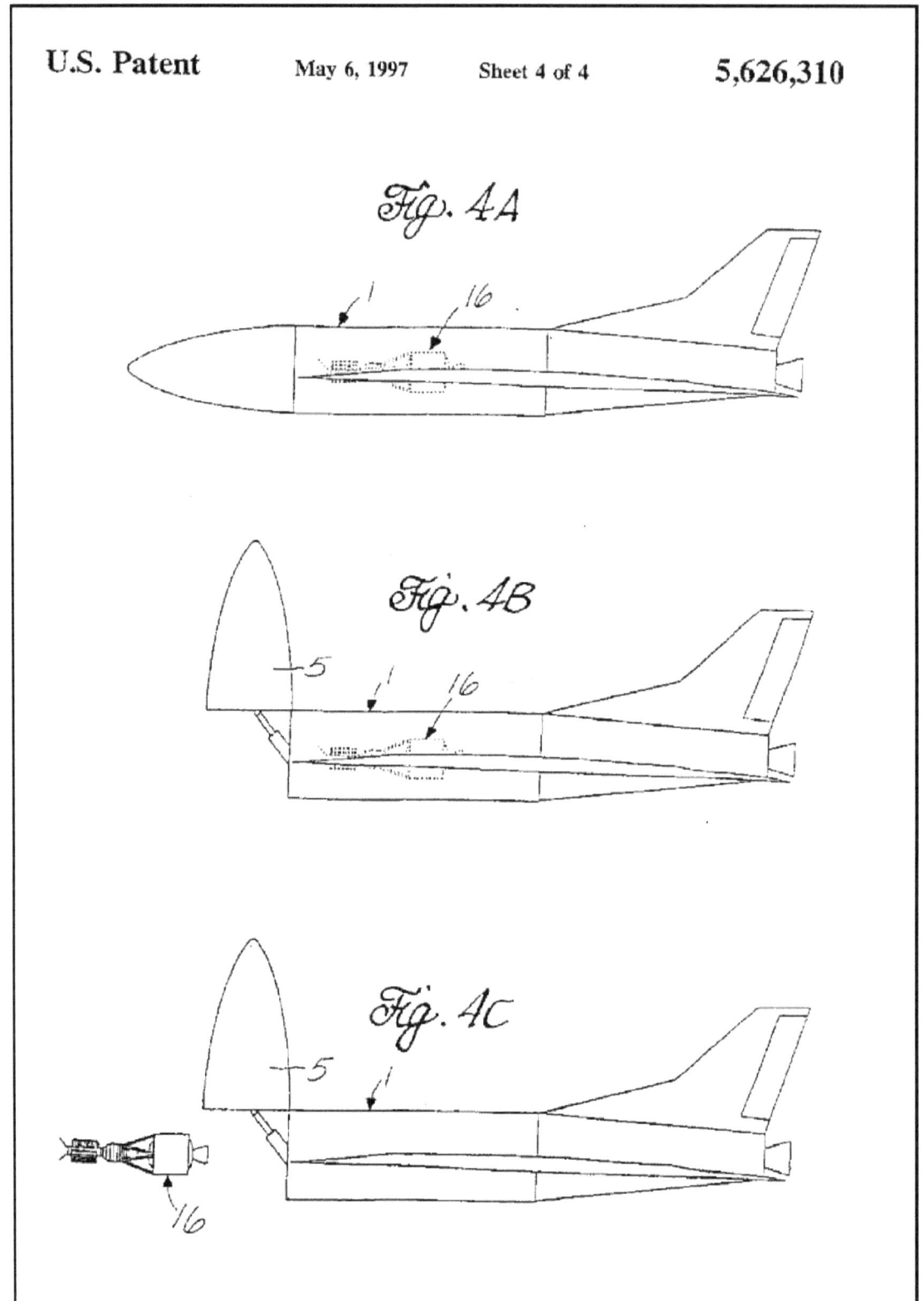

Fig. 4A

Fig. 4B

Fig. 4C

3

propulsion system, effectively offsetting the performance loss usually incurred by the propulsion system having to first offset the vehicle's weight before actually providing acceleration.

The sole current aircraft-launched system (Orbital Sciences Corporation's Pegasus™) has wing surface area only sufficient to partially offset the vehicle's weight at the speed of the launch aircraft. As the vehicle accelerates and, at the same time, becomes lighter by virtue of expending propellant, the wing eventually becomes capable of overcoming the vehicle weight. The performance enhancement potentially available from the wing is hence limited.

The Pegasus™ is carried by its launch aircraft, by direct attachment either to an underwing pylon or a special fitting beneath the aircraft fuselage. Other proposed launch vehicles which are intended to be launched by an aircraft are all designed to be carried by the aircraft in some fashion, either on top of the aircraft, under the wing, or inside the cargo compartment. Some use lifting surfaces, others do not, but in no case is there a design wherein the launch vehicle has aerodynamic lift equal to or greater than the vehicle's launch weight at an indicated airspeed equal to that of the launch aircraft.

Each of these launch vehicles suffers from the same set of deficiencies. First, the maximum weight of the launch vehicle is limited to the weight that the carrier aircraft can safely lift to the required altitude. This places an absolute upper limit on the size and weight of the spacecraft which can be launched by such launch vehicles. The weight limit is not necessarily equal to the cargo capacity of the carrier aircraft. If the launch vehicle is mounted externally to the aircraft, the interference drag added to aircraft by the addition of such appendage will require extra power to overcome. In addition, the structural loads imposed on the aircraft are greater than just the weight of the launch vehicle. The drag force on the launch vehicle and inertial load factors add significantly to the loads applied to the carrier aircraft. A structural limit may be reached long before the actual weight-lifting capacity of the aircraft has been exceeded.

Second, there is risk associated with carrying the launch vehicle, which typically contains large amounts of explosive propellant, on or in a manned launch aircraft. Explosive hazards are reasonably small during flight from the runway to the launch point. The greatest potential for explosion is during or shortly after ignition of the launch vehicle's propulsion system. Partly for this reason, most air-launch concepts require the launch vehicle to fall freely from the carrier aircraft before their propulsion system is started. This reduces the achievable reliability somewhat, in that the launch vehicle is irrevocably separated from its carrier aircraft before it is known with certainty that its propulsion system is functioning properly. There can also be a net loss of performance compared to ground launch if the launch vehicle has no lifting surfaces, and acquires significant speed during free-fall.

Third, the separation of the launch vehicle from the aircraft can introduce dynamic loads to the launch vehicle which are in turn transmitted to the spacecraft. These loads can be very severe, and require a heavier spacecraft structure than might otherwise be needed.

Fourth, externally-carried launch vehicles are subjected to the noise from the carrier aircraft's engines, and to noise generated by the complex air flow around the launch vehicle if it projects into the freestream. This imposes random vibration on the spacecraft. Vibration levels can be higher than those imposed on a spacecraft on a vehicle launched

4

from a ground-based pad, and last hundreds of times longer. Again, a heavier spacecraft structure may be required, and delicate instruments may have to be completely redesigned to survive.

Fifth, the cost and complexity of modifications to the carrier aircraft permitting it to carry the launch vehicle increase dramatically with launch vehicle size. In fact, such modifications may become more complex and expensive than building a launch pad, reducing the incentive to utilize aircraft launch.

Finally, there is a risk to the aircraft crew from a multitude of failures which can occur when separating a launch vehicle from the aircraft. As one example, the launch vehicle control system may fail resulting in collision with the carrier aircraft and loss of both.

While launching of space launch vehicles from aircraft has significant advantages over ground-launch, the limitations associated with current designs are significant. Most important is the limitation on spacecraft size and weight imposed by current technology. In order to more fully realize the advantages of aircraft launch of space launch vehicles, as well as reduce its cost, risks, and other limitations, a new approach is desired.

SUMMARY OF THE INVENTION

The present invention overcomes the deficiencies of current aircraft-launched space launch vehicle technology through the application of glider technology to the launch vehicle. Simply stated, this consists of adding lifting surfaces to the launch vehicle which are capable of overcoming the vehicle's launch weight at speeds less than or equal to the takeoff speed of a conventional aircraft. The launch vehicle may then be towed, using a flexible cable, behind a conventional aircraft. The launch vehicle, in tow, can then be flown to any desired geographic location in exactly the same manner as a launch vehicle carried on or inside of an aircraft. At the launch point, the tow line can be released and the launch vehicle's propulsion system started in a safe, stable manner, and propel the vehicle's payload into orbit.

The invention consists either of a glider airframe with one or more propulsive stages incorporated into it, or alternatively of a launch vehicle of one or more propulsive stages to which suitable lifting surfaces have been appended. The vehicle can be either completely expendable, partially reusable, or completely reusable depending on the specific vehicle requirements. It may be equipped with landing gear in order to permit it to be recovered in the event of inability to launch. In any embodiment, it is equipped with attachment points and release mechanisms for the tow line, and a control system which permits it fly either autonomously or under remote control.

Ground handling and takeoff would be accomplished by mounting the vehicle on a carriage equipped with wheels and a braking system capable of stopping the vehicle safely in the event of an aborted takeoff. The carriage would be left on the ground to save weight, and would use its integral braking system to stop automatically once the launch vehicle has lifted off.

The tow aircraft contributes only thrust, not lift, to the launch vehicle. The total engine thrust available from a commercial wide-body transport jet's engines is far in excess of the aircraft's drag. The difference between engine thrust an aircraft drag can be directly applied to the launch vehicle, which reacts the applied load with its own drag force. The maximum weight of the glider is then limited only by its lift-to-drag ratio (L/D), and is roughly equal to the applied tow load multiplied by the L/D. To give a specific example, the 747-200B at cruise may have a total available thrust of 67,500 pounds force at 36,000 feet cruise altitude. If the 747 weighs 500,000 pounds, and its L/D is 12, the drag force on it at equilibrium cruise is 41,667 pounds, leaving 25,900 pounds net force to apply to the tow cable. If the launch vehicle has an L/D of 10, its maximum weight can then be 259,000 pounds. By contrast, the maximum weight which can be carried on the aircraft's available structural hard points (which are used for transporting spare engines) is 50,000 pounds.

An implication of the above is that far fewer structural modifications need to be made to an aircraft to enable it to tow a heavy load than to carry a light load. In the example given, a 259,000 pound launch vehicle could be towed behind an aircraft and exert a force on the aircraft of only 25,900 pounds. Yet existing hard points on the aircraft are already capable of reacting 50,000 pounds of force.

From the perspective of the launch vehicle, being towed relieves it of the need to carry heavy propulsion systems or fuel to carry it from the runway to the point of powered boost ascent. This simplifies the launch vehicle, and effectively transfers the burden of getting from the takeoff point to the point of powered ascent initiation to the tow aircraft in the same manner as a launch vehicle carried aboard an aircraft.

Since the launch vehicle is equipped with wings which permit it to take off at aircraft speeds, it could obviously take off from the ground under its own power. Its performance would be reduced, however, since it would have to overcome more drag, gravity, and back-pressure losses. This is an instance where the use of the term "zero-stage" for the tow aircraft is appropriate. The method of attaching the launch vehicle and zero-stage together, via flexible cable, is made possible by the aerodynamic lift capability of the launch vehicle, and constitutes a significant advance in the state of the art for launch vehicles.

Other advantages accrue from the use of high aerodynamic lift of the type described above. The use of high-lift devices in launch vehicles permits them to perform in a manner not possible to low-lift vehicles such as Pegasus™. Low-lift vehicles must have high thrust in order to minimize their performance loss due to overcoming gravity. For a given amount of propellant, the duration of thrust is inversely proportional to the thrust level. High thrust means short burn times, which cause the vehicle to reach relatively high speeds at relatively low altitudes. This imposes a performance loss due to drag that would not otherwise occur.

A high-lift vehicle can climb at a shallower angle for a longer period of time, since it is supported entirely aerodynamically. Thrust is required only to acquire or maintain speed, unlike the case of a low-lift vehicle which requires significant additional thrust to offset the vehicle's weight. The high-lift vehicle, burning the same amount of propellant, can climb to higher altitude before acquiring significant speed than can a low-lift vehicle, reducing the drag penalty. Having such significant force available on demand can also aid in shaping the trajectory to minimize gravity losses, and even in changing the flight azimuth after significant speed has been acquired, without an attendant loss of performance.

At the end of the flight, if the glider is to be recovered, high-lift can work to reduce heat loading on the vehicle, and extend its range. Once the first-stage propellant has been expended, the vehicle's wing loading is so low compared to its takeoff value that heating and maneuvering loads are much more benign than would be possible with a low-lift vehicle.

The high-lift aspect intrinsic to the towed-glider launch vehicle sets it apart from all other air-launched or aerodynamically assisted concepts. It operates in a different flight regime than low-lift vehicles, one that has several advantages. There are also numerous practical benefits which accrue from the towed-glider launch vehicle, such as the above-mentioned simplification of aircraft modifications.

BRIEF DESCRIPTION OF THE DRAWINGS

FIGS. 1-A through 1-D show one embodiment of a launch vehicle of the type described herein. FIG. 1-A is a platform view, FIG. 1-B a side view, and FIGS. 1-C and 1-D are side views illustrating the operation of an articulating nose door.

FIGS. 2-A through 2-D illustrate the layout of the propulsion systems in this embodiment of the launch vehicle.

FIGS. 3-A through 3-D illustrate the takeoff sequence for the launch vehicle and tow aircraft.

FIGS. 4-A through 4-C illustrate the method of separating the upper stages from the first stage in this embodiment of the latch vehicle.

DETAILED DESCRIPTION OF THE INVENTION

One embodiment of the invention, in FIG. 1, shows a glider airframe [1] equipped with wings [2] and rudder [3], into which a rocket propulsion system is incorporated, as indicated by the nozzle [4] projecting from the aft end. FIG. 1-A is a platform view, showing a cranked-delta wing configuration. This wing configuration was chosen to give an optimum balance between subsonic lift-to-drag ratio and hypersonic drag, allowing the maximum weight to be towed behind a conventional aircraft while imposing the least drag penalty at high speeds. Other planforms, including variable-sweep and X-wing configurations would be equally suitable.

FIG. 1-B shows the vehicle in side view, with the rudder [3] more clearly indicated. FIGS. 1-C and 1-D illustrate one possible implementation of a means of loading and deploying the upper stages and spacecraft, through the use of an articulating nose door [5]. This arrangement is similar to cargo doors on conventional nose-loading freighter aircraft, such as the 747-100F and the C-5A Galaxy. FIG. 1-C shows the door partially opened, and FIG. 1-D shows it fully opened.

FIG. 2-A shows the vehicle in section, illustrating the integrated propulsion system [6 through 9], the bay for upper stages and spacecraft [10], and a structural interface for the upper stages and payload [11]. The tank located in the nose [6] would hold liquid oxygen (LOX) in this embodiment, as would the aftmost tank [8]. The center tank [7] would hold kerosene. This arrangement was chosen to permit transfer of propellant along the length of the vehicle in such a manner as to keep the vehicle center of gravity ahead of its center of pressure through all flight regimes. During the transition from subsonic to supersonic flight, the center of pressure moves forward significantly. If the center of gravity is not kept in a certain relation to the center of pressure, the vehicle becomes unstable. By depleting the LOX in tank [8] first, the center of gravity can be made to travel forward as propellant is expended. In an abort situation, wherein the engine [9] shuts down, the vehicle will decelerate. The accompanying aftward shift in the center of pressure location can be compensated by transferring residual LOX from the forward tank [6] to the aft tank [8], thus maintaining a stable relationship of center of pressure and center of gravity locations.

FIG. 2-B shows the components of the upper stage and spacecraft assembly. A large solid propellant motor [12] serves as the second stage of the launch vehicle. A small solid propellant motor [13] serves as the third and final stage. The two motors are joined by a truss or other structural assembly [14]. A spacecraft [15] can then be joined to the third stage, resulting in the integrated spacecraft and upper stage assembly [16] shown in FIG. 2-C. This assembly is then installed in the first stage as shown in FIG. 2-D. During ground operations, the integrated assembly of upper stages and spacecraft can be loaded into the launch vehicle horizontally, through the open nose door, eliminating the need for cranes or other heavy-lift equipment normally associated with pad-launched launch vehicles. This represents a considerable saving in equipment cost, and in the complexity and time required to perform pre-flight assembly. Since the liquid propellants for the launch vehicle would not be loaded until just before takeoff, the nose tank [6] will be empty during the loading operation of the spacecraft and upper stage assembly [16], so that the hinge structure and opening mechanisms need not be excessively strong and heavy.

All ground operations would be performed with the launch vehicle mounted to its handling and takeoff cart [17], as shown in FIGS. 3-A through 3-D. Mechanical attachment of the vehicle [1] to the cart [17] would be accomplished by the use of explosive bolts, or some other mechanism which would securely fasten the two together, yet which could be released on command. The launch vehicle would be coupled to the tow aircraft by a flexible cable [18]. This cable would have suitable attachment and release mechanisms located on the launch vehicle [1], and would be attached to the tow aircraft [19] through a winch mechanism mounted in a fairing [20] at or near the tow aircraft's center of gravity. This is done to minimize the overturning moments which would be applied to the aircraft by the tow line.

FIG. 3-A shows the assembly during takeoff roll. Both vehicles remain on the ground until the tow aircraft has passed its rotation speed, which is the speed needed to take off. Current flight practices required jet aircraft to take off after this speed has been reached, even if a serious mechanical problem arises with the aircraft. At this point, as shown in FIG. 3-B, a hydraulic ram [21] on the carriage extends to lift the nose of the launch vehicle to its takeoff angle. The mechanical linkage between the carriage and the launch vehicle is then severed, and the launch vehicle takes off as shown in FIG. 3-C. When the launch vehicle has reached a suitable altitude, the tow aircraft can then rotate for take off as shown in FIG. 3-D.

There are two reasons for this takeoff procedure. First is that once the launch vehicle becomes airborne, the tow aircraft must also take off even if it has developed a problem which will not permit it to continue the mission. In such a situation, propellant can be jettisoned rapidly from the launch vehicle to lighten its weight for subsequent recovery. The tow aircraft can execute a turn to bring it back to the runway for emergency landing, and the two vehicles can be recovered without incident for future flight attempts.

The second reason for having the launch vehicle airborne first is to ensure that it is out of reach of the strong wing-tip vortices which develop when a large aircraft takes off, or otherwise flies at a high angle of attack. During ascent to the launch point, the launch vehicle continues to fly above the tow aircraft to avoid these vortices.

The launch vehicle is towed to a desired launch location, during which transport time the necessary preflight

checks are performed telemetrically through a launch console located in the tow aircraft. The launch vehicle is also piloted remotely, by a pilot located in the launch aircraft and using standard Remotely Piloted Vehicle (RPV) control technologies. Once at the desired launch location, the first stage rocket engine is ignited, and once its operation has been verified, the tow line is cast off from the launch vehicle.

The launch vehicle climbs to a suitable altitude and velocity, then enters coasting flight. In the embodiment shown, the first stage propellants are exhausted at an altitude of approximately 350,000 feet and a velocity of 14,000 feet per second. The flight path angle at first stage shutdown is such that it can coast to 600,000 feet or more. Once the vehicle has coasted above 400,000 feet, it is out of the sensible atmosphere. Aerodynamic forces and free molecular heating are no longer a concern, and the articulating nose door may be opened for deployment of the spacecraft and upper stage assembly.

FIGS. 4-A through 4-C illustrates separation of the first stage from the spacecraft and upper stage assembly during flight. In FIG. 4-A, the vehicle is in coasting flight. In FIG. 4-B, the nose door is shown in the open position. In FIG. 4-C, the spacecraft and upper stage assembly is shown after being ejected from the first stage. This can be accomplished using qualified spring separation mechanisms, hydraulic rams, or other suitable actuators.

Once separated, the spacecraft and upper stage assembly coasts to a distance from the first stage to avoid damage to the latter from jet impingement. The second and third stage motors then fire in sequence to place the spacecraft into orbit. The door on the glider would then be closed, and the glider would reenter the atmosphere for subsequent gliding flight to a recovery landing field.

This is the preferred embodiment for initial development, because it represents the most cost-effective solution in terms of initial and operational cost. Commercially available expendable upper stages may be used, requiring no development cost. The recoverable rocket-propelled glider is readily developed using existing airframe and propulsion technologies. Guidance and navigation systems are commercially available for controlling the vehicle through all flight regimes, including automated landing of the first stage and orbital injection of the third stage.

A liquid propulsion system is preferred in a recoverable rocket, since it is more readily refurbished and refuelled than either a solid or hybrid rocket system. System safety is also enhanced, since in an aborted flight situation, liquid propellants can be jettisoned from the vehicle to lighten it for landing and reduce explosive hazard. This cannot be done with solid propellant motors, and is only partly possible with hybrids. However, the invention does not depend upon any specific propulsion technology. Its advantages are independent of the types of propulsion systems used, and selection of the types of systems need depend only on a given set of requirements. In the preferred embodiment, the primary objective is to reduce the cost of space launch, and the selection of propulsion systems reflects that fundamental objective.

An advantage of liquid or hybrid propulsion systems is the ability to vary thrust level at will. This permits taking full advantage of the ability to minimize gravity losses by climbing at shallow angles for extended periods of time, in that throttling back the engine or engines conserves propellant. Bipropellant liquid propulsion systems entail additional safety risk compared to hybrids due to the presence of two liquids. However, the tankage for liquid propulsion systems can be distributed through the launch vehicle in a manner which makes best use of available volume, and permits control of the location of the vehicle center of gravity.

During tow, the distance between the launch vehicle and tow aircraft can be varied using a winch mechanism. By controlling the separation of the tow aircraft and launch vehicle, random vibration imposed on the spacecraft from the aircraft engine noise and aerodynamic buffeting from the tow aircraft wake can be minimized. This is in sharp contrast to other external-carry air launch concepts, in which engine and aerodynamic noise can impose more severe vibration environments on the spacecraft than the reflected rocket noise of a launch vehicle as it takes off from a ground-based pad.

The launch vehicle can also be positioned far enough behind and above the tow aircraft to permit ignition of the launch vehicle's propulsion system while the tow line is still connected, without endangering the crew of the tow aircraft. This provides enhanced reliability for the launch system, since proper operation of the launch vehicle's engine can be verified prior to irrevocable severing of the tow line. If the launch vehicle's propulsion system fails to start properly, it can be shut down and the tow aircraft and launch vehicle returned to the launch site safely. Even in the event of a catastrophic failure of the launch vehicle upon propulsion system ignition, the tow aircraft can be far enough away to prevent damage from explosive overpressure or shrapnel impact. The fact that the relative wind blows from the tow aircraft toward the launch vehicle at hundreds of miles per hour enhances the safety of the tow aircraft.

Use of expendable upper stages simplifies the development of the vehicle in this embodiment in other respects. By placing half of the propulsive burden on motors which are commercially available, it requires no extensive development of upper stages. More importantly, however, it simplifies the task of protecting the recoverable first stage from aerodynamic heating during ascent and, especially, during reentry.

During ascent, the rocket-powered glider does not achieve sufficient velocity within the sensible atmosphere to make aerodynamic heating an intractable problem. Use of throttling in the first stage propulsion system simplifies the problem further, since low speeds can be maintained without penalty for extended periods of time. This allows the vehicle to climb to a sufficient altitude to permit it to throttle up and "dash" through the hypersonic portion of flight in a relatively short time.

Reentry heating is significantly less for this glider than that experienced by a vehicle entering the atmosphere from orbit, for two reasons. The first is that the maximum velocity of the first stage need never exceed half of that required to achieve orbit. This in turn means that the vehicle has to dissipate no more than 25% of the energy possessed by an orbiting body in order to slow down to subsonic flight speed. Also, the weight of the glider on takeoff must be between three and five times that of its weight after expending its propellant. The wing-loading of the glider is thus one-third to one-fifth its takeoff value. This permits energy to be dissipated over a larger area, resulting in lower heat transfer rates to the vehicle structure. Heating loads may thus be accommodated by application of simple, durable insulation materials over most of the structure, and refractory materials in stagnation regions.

Overall, this embodiment represents the best balance of development cost and risk and operational cost and risk of any near-term system whose primary objective is to mini-

mize cost and risk. Other implementations are possible, employing other types of propulsion systems, including airbreathing systems, in the glider, and recoverable upper stages. The embodiment described herein is preferred mainly due to the fact that it does not tax the state of the art in aircraft or launch vehicles, but combines elements of both in a simple fashion which nonetheless results in a significant advance in the state of the art.

I claim:

1. A towed glider space launch vehicle adapted to be towed by an aircraft comprising:
   a first vehicle having
      aerodynamic lifting surfaces providing lift sufficient to support atmospheric flight of the first vehicle at an airspeed less than the takeoff speed of a conventional aircraft,
      an integral payload bay,
      access means for ingress and egress from the payload bay,
      means for releasable attachment of a tow cable, and
      a throttleable rocket propulsion engine for increasing the velocity of the first vehicle;
   means for receiving a spacecraft in the payload bay through said access means, said spacecraft ejectable through said access means during flight of the first vehicle.

2. A space launch vehicle as defined in claim 1 further comprising:
   an upper stage propulsion system having an interface for attachment of the spacecraft, said upper stage propulsion system received in the payload bay through said access means and ejectable through said access means during flight of the first vehicle.

3. A towed glider space launch system adapted to be towed by a conventional aircraft comprising:
   a first vehicle having
      aerodynamic lifting surfaces providing lift sufficient to support atmospheric flight of the first vehicle at an airspeed less than the takeoff speed of a conventional aircraft,
      an integral payload bay,
      access means for ingress and egress from the payload bay,
      means for releasable attachment of a tow cable, and
      a throttleable rocket propulsion engine for increasing the velocity of the first vehicle;
   a second vehicle received in the payload bay through said access means and having
      an interface for attachment of a spacecraft, and
      an upper stage propulsion system, said second vehicle ejectable through said access means.

4. A space launch system as defined in claim 3 wherein the first vehicle includes a fuselage portion incorporating the integral payload bay and the access means comprises:
   a nose section adapted for closure of a main body portion of the fuselage containing the integral payload bay; and
   means for articulating the nose portion between a first closed position and a second open position, said second open position exposing the spacecraft for ejection from the payload bay.

5. A space launch system as defined in claim 3 wherein the first vehicle includes a bi-propellant tankage system for the throttleable rocket propulsion engine, the tankage system including a first forward tank and a second aft tank, said forward and aft tanks interconnected for transfer of fluid to control position of a center of gravity for the vehicle with respect to a center of pressure produced by the aerodynamic surfaces.

11

6. A space launch system as defined in claim 5 wherein the launch carriage further incorporates means for positioning the first vehicle in a first horizontal position and in a second position at a takeoff angle.

7. A space launch system as defined in claim 3 further comprising a launch carriage on which the first vehicle is severably mounted, said carriage incorporating a plurality of wheels for rolling takeoff of the first vehicle under tow by a launch aircraft, said carriage severed from the first vehicle upon liftoff.

8. A method for launch of a spacecraft employing a first launch vehicle having aerodynamic lifting surfaces providing lift sufficient to support atmospheric flight of the first vehicle at an air speed less than the takeoff speed of a conventional aircraft, an integral payload bay, access means for ingress and egress from the payload bay, means for releasable attachment of a tow cable, and throttleable rocket propulsion engine, a second vehicle received in the payload bay through said access means and having an interface for attachment of the spacecraft and an upper stage propulsion system and a tow aircraft, the method comprising the steps of:

inserting the second vehicle in the payload bay of the first vehicle;

attaching a tow cable from the tow aircraft to the first vehicle;

accelerating the tow aircraft past its rotation speed on a runway;

controlling the first vehicle for takeoff;

rotating the tow aircraft for takeoff upon the first launch vehicle attaining a suitable altitude;

flying the tow aircraft to a desired launch location;

igniting the rocket engine of the launch vehicle;

casting off the tow line from the launch vehicle upon verification of proper rocket propulsion operation;

controlling the launch vehicle for climb to a predetermined altitude and velocity;

opening the access means to the payload bay;

ejecting the second vehicle from the payload bay;

operating the upper stage propulsion system for insertion of the spacecraft into orbit;

closing the access means to the payload bay; and

controlling the first vehicle for atmospheric reentry and gliding flight to a recovery landing field.

9. A method as defined in claim 8 wherein the first launch vehicle includes a liquid bi-propellant system for the throttleable rocket propulsion engine, said propellant system incorporating a forward tank and an aft tank interconnected for transfer of fluid, and the step of controlling the launch vehicle further comprises the step of regulating extraction of propellant from the forward and aft tanks to achieve a forward shift of the center of gravity relative to the center of pressure from the lifting surfaces to accommodate transition from subsonic to supersonic flight.

10. A method as defined in claim 8 further comprising of the steps of transferring propellant between the forward and aft tank to control the center of gravity of the vehicle in relationship to the center of pressure of the aerodynamic lifting surfaces to accommodate varying flight velocity.

11. A method as defined in claim 8 wherein the step of flying the tow aircraft further comprises the step of varying the distance between the tow aircraft and the first launch vehicle to minimize vibration on the launch vehicle and

12

spacecraft from engine noise of the tow aircraft and aerodynamic buffeting of the tow aircraft wake.

12. A method as defined in claim 8 wherein the step of flying the tow aircraft further comprises the step of adjusting the position of the launch vehicle in relation to the tow aircraft to permit ignition of the launch vehicle's propulsion system while the tow line remains connected without endangering the tow aircraft.

13. A method as defined in claim 12, including a procedure for maintaining the first launch vehicle under tow after a system failure during the step of igniting the rocket engine comprising the steps of:

detecting a launch vehicle malfunction;

shutting down the launch vehicle rocket propulsion engine; and

maintaining the launch vehicle in tow for return to a landing site.

14. A method as defined in claim 8 further incorporating a procedure for emergency return to launch site comprising the steps of:

detecting a mission abort condition in the tow aircraft or launch vehicle;

jettisoning propellant from the launch vehicle to lighten its weight for subsequent recovery;

maneuvering the tow aircraft for return to the runway for emergency landing; and

recovering the tow aircraft and launch vehicle by conventional landing.

15. A method as defined in claim 8 wherein the step of controlling the launch vehicle includes the steps of:

throttling the launch vehicle propulsion system for a predetermined ascent profile to maintain a predetermined aerodynamic heating level; and

throttling up the rocket propulsion engine for a dash to a final hypersonic velocity prior to ejection of the second vehicle.

16. A towed glider space launch vehicle system comprising:

A tow aircraft adapted to tow a glider;

a glider having

aerodynamic lifting surfaces providing lift sufficient to support atmospheric flight of the glider at an airspeed less than the takeoff speed of the tow aircraft,

an integral payload bay in the glider.

access means for ingress and egress from the payload bay.

means for releasable attachment of a tow cable between the tow aircraft and the glider, and

a throttleable rocket propulsion engine for increasing the velocity of the glider.

means for receiving a second vehicle in the payload bay through said access means.

said second vehicle having

an interface attachment for a spacecraft, and

an upper stage propulsion system, said second vehicle being ejectable through said access means for launch during flight of the glider.

17. A system as defined in claim 16 wherein the throttleable rocket propulsion engine is adapted to lift the glider above the sensible atmosphere for launch of the second vehicle.

* * * * *

Eclipse
EXD-01 Flight 5
First Tethered/Release Flight
December 20, 1997

Project Manager's Comments

I have never experienced a real first research flight before. There have been a few interim mods and the like, but never anything unique like Eclipse. And I have to say, it was a very big thrill for me!

That statement is obviously based upon the fact that I felt very confident that the project team was ready to go. And it was.

Day of flight was another very, very cold day which give us our share of problems. Conducting the mission on a Saturday gave use the flexibility to focus on problem resolution and not feel the pressure of time. There was enough tension (not tow rope tension) in the sequence of events, hydraulic leaks and chase aircraft fuel problems, to name a few. But in the end, we accomplished a highly successful and historical event.

In addition, we have completed two out of the three project objectives. 1) Towing a delta wing airplane with a transport type airplane. 2) Towed flight operations and procedures. The remaining objective is to validate the simulation.

As with most research efforts we were surprised by a few things. The tow rope flailed wildly after release and whipped off the knuckle by breaking the frangible link. And the behavior of the tow rope is not a straight line element as was predicted. And an inadvertent tow rope release. But -- the tow rope tension was as predicted. Procedures were flawless used properly. Data systems worked as planned. Data processing was timely. This was an EXCELLENT first flight.

And the timing wasn't entirely an unhappy circumstance. True, it would have been very nice to have achieved our first flight earlier in the year. But flying before the end of the year was an acceptable substitute.

From this perspective, it looks to me like Eclipse is really going to be fun and productive!

Carol A. Reukauf

Document 3. Eclipse EXD-01 Flight 5, First Tethered/Release Flight, 20 December 1997, Project Manager's Comments, Carol A. Reukauf

Eclipse
Pilot's Flight Test Report
EXD-01 Flight 5 - First Tethered Flight

Pre-start
The EXD pilot arrived at the aircraft at the proper time for scheduled pilot entry to find the two chase F-18s were still awaiting fueling (C 1)[1]. MCC recommended that both test aircraft delay their engine start until the fueling was complete. This accounted for approximately a 30 minute delay. The chase planes were not refueled the day prior because Maintenance felt there was the potential for the test flight to not occur, which could cause the fueled aircraft to leak in the hangar over the Christmas holidays. **Recommend additional coordination or emphasis on prompt refueling be completed prior to future early morning operations (R1)[2].**

During the fuel delay the pilot decided to complete the EXD preflight. Inspection revealed the pneumatic system had less than 2000 PSI. The pilot questioned the crew chief who responded it had recently been charged to 3000 psi, and the drop in pressure was due to the 20°F temperature. He further stated the nitrogen cylinders, which were used to charge the aircraft, had dropped 600 psi themselves in the short time they had been outside. Attempts to put additional pressure into the system failed.

The pilot was skeptical the engine would start and decided he should not wait for the second chase aircraft to complete refueling but to go ahead with engine start. As was decided after the high speed taxi test, to avoid having to use the cold weather sensitive MC-11 start cart an internal air start was to be attempted, and pneumatic cylinders would then recharge the internal bottles.

Start
The pilot attempted engine start and the engine accelerated to 20% RPM but failed to ignite. The pilot cleared the engine and the crew was forced to again try to recharge the pneumatics. This took repeated attempts with various cylinders before the aircraft began taking air. Unfortunately, the MC-11 was not prepositioned in the event of a pneumatic recharging problem.
Recommend in the future we always have the MC-11 in position to aid in pneumatic charging (R2). The crew suspected internal icing was inhibiting pneumatic system recharging of the EXD (C2).
Eventually the crew was able to get nitrogen flow into the EXD, and it was recharged to 2700 psi. By this time the entire operation was nearly one hour behind the planned timeline.

Post Start and Taxi Onto the Runway
The second start attempt went well with the usual extremely slow spool-up of the J-75 during cold weather operations with JP-8 fuel. There were no issues with the C-141A which had already started up and was waiting patiently in the EOR.

---

[1] Numbers preceded by a "C" indicate a conclusion.
[2] Numbers preceded by a "R" indicate a recommendation.

Document 4. Eclipse Pilot's Flight Test Report, EXD-01 Flight 5—First Tethered Flight, Mark P. Stucky, Eclipse Project Pilot

Eclipse
Pilot's Flight Test Report
EXD-01 Flight 5 - First Tethered Flight

The taxi onto the runway went better than that of the high-speed taxi test, but the EXD pilot felt he was led too far down the runway and the truck driver felt he should drive referencing the EXD instead of as the "flight lead" (C3). Since there have been no unexpected rope tensioning issues during the taxi onto the runway, recommend using the procedures used during the CST (truck driver leads the EXD but references the aircraft for his positioning) (R3). Additionally, the tow rope was not anchored to the base of the pitot boom (as had been done previously) and the crewman handling the rope allowed the knuckle assembly to slam against the stops (C4). This slamming could potentially damage the cable positioning transducers. Recommend the tow rope be anchored to the base of the pitot tube until the EXD is in position and hold (R4).

Hook-up
The rope hookup was uneventful, although some relaying of information was required between the mobile and the C-141A.

Slack Removal & Tensioning
A new person was used to hold the rope staff, and this required more supervision by the crew chief and diverted some of the attention of the pilot during the slack removal. By chance, the C-141A Loadmaster was also distracted as the rope pulled tight, and no ICS or UHF calls were made. The EXD pilot noticed the rope pulling tight and transmitted "hold your position" as the tension quickly rose to 4000 lbs. Inattention during slack removal could cause high tension values and/or movement of the EXD which could be a hazard to the rope handler (C5). The EXD pilot feels that once the aircraft is in position on the runway, the lateral offset will preclude the rope from pulling into the pitot tube and therefore the rope handler should no longer be required. Recommend that once in position and hold, the rope handler remove the anchor at the nose and move to the side of the aircraft where he would be clear of any hazard but still available to assist if required (R5).

A big surprise occurred when the crew chief returned to the EXD, pointed at the nose gear and gave an emphatic thumbs down. He got on the mobile radio and informed the team the nose wheel steering system was leaking hydraulic fluid. The pilot had engaged the nose wheel steering system for the taxi and only a single hard turn to align had been accomplished. He disengaged the system, and had the crew chief reinspect. The crew chief indicated the leak had stopped, so he signalled for the pilot to exercise the system. Repeated full commands were done without further leaking. The most likely cause of the nose wheel steering leak was the failure of a seal to initially seat due to the cold weather (C6). The team was once again ready to proceed.

The photo chase tookoff early to fly the planned routing while checking for turbulence. The video chase was tasked to do an airborne pickup and tookoff just prior to the rope hookup. The chase's decision to takeoff early

Eclipse
Pilot's Flight Test Report
EXD-01 Flight 5 - First Tethered Flight

avoided any interference with the runway FOD check and allowed time for both him and the video operator to practice positioning (C7). Recommend future video chase pilots be briefed to takeoff just prior to the tow rope hookup (R6).

Tensioning was quickly accomplished, and the team made the "go for flight" calls.

Tethered Takeoff
When the tower gave clearance for takeoff the chase pilot was already at the midfield point and radioed he would do another lap. The EXD pilot requested he make it a short lap because he was holding the brakes under tension. Regardless, the call for brake release came nearly three minutes after clearance for takeoff which the EXD pilot felt was excessive (C8). Recommend the EXD pilot ensures the airborne pickup pilot knows the importance of anticipating the clearance for takeoff and avoiding any unneccesary delays (R7).

The pilot anticipated the long delay from the brake release call and allowed the tension to build to 10,000 lbs. prior to easing the brake pressure. He continued to ease up on the pedals as the EXD built up speed. Within approximately ten seconds of roll he was able to completely release any brake pressure and the acceleration maintained the tension above 6000 lbs. There were no obvious large tension oscillations during the ground roll and the tow rope remained off the runway at all times (C9). Recommend the same brake release procedures be used for future takeoffs (R8).

During the ground roll the pilot noticed it took rudder (nose wheel steering) input to maintain the lateral offset from the C-141A. The C-141A rotated and tookoff as planned, and the EXD pilot allowed the lateral offset to null out as he rotated. The pilot estimated the C-141A to be 200 ft. AGL but no UHF transmissions were heard. The EXD began getting light on its wheels around 155 KIAS. The pilot now knew without a doubt the C-141A had to be above 200 ft., and he was now starting to apply forward stick to keep the EXD from getting airborne. The EXD felt stable and a quick check of the airspeed indicated 170 KIAS and accelerating. The EXD pilot realized the C-141A was not transmitting the planned descriptive calls and elected to allow the EXD to fly.

Takeoff occurred just prior to the midfield taxiway which matched well with predictions and a five knot tailwind. The video chase was in proper position at all times.

The pilot transmitted "Eclipse is airborne" and after several seconds radioed the tow was stable. The C-141A's check-in with Sport radar was the first transmision that either the EXD or MCC had heard from them since the beginning of the takeoff roll.

Eclipse
Pilot's Flight Test Report
EXD-01 Flight 5 - First Tethered Flight

The pilot estimated he was near the lower tow limit and climbed several degrees while trimming a few clicks of nose down. The pilot radioed the tow was extremely low gain, and in fact was stable enough that no pilot input whatsoever was required. The flight was over 2000 ft. AGL by the east lake shore and commenced the turn to the north. As per the simulation the EXD tracked nicely into, through, and out of the turn without any pilot control input.

Preliminary indications (EXD declination @ 170 KIAS and C-141A aircrew comments) show that the **170 KIAS takeoff speed was acheived just prior to the C-141A reaching 350 ft. AGL (C10). Recommend reducing the C-141A rotation and initial pitch attitude to +4° to decrease the altitude at which 170 KIAS is reached (R9).**

Flight Cards
Tethered Stick Raps During Climbout. The raps were done with no adverse reactions noted.

Tethered Doublets During Climbout. From an estimated nominal tow position of -12° the pilot executed small longitudinal and lateral/directional doublets. The EXD was stable although lightly damped.

A small area of light turbulence was encountered which was reported by both the C-141A as well as the photo chase F-18. The EXD pilot did not feel the bumps of the turbulence, but noted an excitation of dutch roll similar to that seen in the pilot-in-the-loop simulator during turbulence encounters.

The doublets were repeated at approximately -8° and -16° elevation with similar effects. The flight leveled off at 10,000 ft. MSL prior to initiating the lateral offset doublets.

While leveling and turning to the east the EXD pilot observed the canvas covered midpoint of the tow rope to be flapping like a flag in the breeze. After a short period of time the rear of the canvas began splitting and flapping violently. The pilot anticipated departure of parts of the canvas and postponed further test cards until it occurred. Several feet of canvas were immediately shed and passed well above the EXD.

From the EXD pilot's point-of-view, the canvased area was more lively than the rest of the rope. While the Vectran often appeared virtually motionless, the canvas sleeve was always moving at a high frequency. This could have been caused by the increased drag of the large canvas sleeve. **The canvas sleeving was not required to protect the nylon strapping during ground operations (C11). Recommend consideration be made to eliminate the canvas sleeve entirely or limit it to two small sleeves around each of the two-pin connectors (R10).**

Tethered Doublets Dirty @ 10K ft MSL. This series of doublets felt similar to those during climbout except the EXD felt looser or less damped at the upper

Eclipse
Pilot's Flight Test Report
EXD-01 Flight 5 - First Tethered Flight

and lower positions (approximately -10° and -16°, respectively). In these positions the EXD pilot felt he had to actually fly the aircraft to keep it steady, although it could still be flown hands off although with greater deviations. The pilot noted the doublet induced oscillations could be seen in the tow rope more readily than they could be felt in the aircraft, especially after several cycles.

At the upper position the tow rope appeared nearly straight which made the EXD pilot uncomfortable due to the decreased vertical separation between the remaining canvas and the EXD. In the lower position a very obvious rope catenary was noted. From the EXD pilot's perspective most of the catenary seemed to occur on the lower half of the rope.

The flight turned south, and an angle of bank of approximately 25° was reached. The EXD remained stable throughout the turn although the pilot noted aft stick pressure was required to maintain the declination angle. **The EXD was stable in small and moderately banked turns (C12). Recommend future test points be completed to investigate the ability for tethered flight at high bank angles (R11).**

The pilot set up for the lateral offset and, as in the simulator, moving laterally felt like trying to move a brick wall and only a few degrees of offset was accomplished. The pilot noted he was approximately behind the C-141A's number 4 engine.

As expected, the pitch doublet caused a coupling in the lateral/directional axis. The pilot was steadying the EXD for the lateral/directional doublet when the tow rope was suddenly released under approximately 6000 lbs. of tension.

The knuckle assembly immediately sprang well clear of the EXD, and the tow rope began whipping vertically. The EXD pilot radioed that a release had occurred and cautioned the chase aircraft to remain clear of the whipping rope. On the second hard whip, the knuckle assembly departed the tow rope and fell to the desert below. **Separation of the knuckle assembly should be an anticipated consequence of any tow rope release (C13).** Since controllability during tethered turns is not an issue, **recommend future operations immediately occur in the PIRA airspace over uninhabited areas (R12).**

The release of the EXD was a complete "non-event" and did not require any immediate action. There was not any dramatic deceleration nor any tendency to overrun the C-141A. After the knuckle fell away, the EXD pilot banked the aircraft and started to follow it down until he was reminded by the test conducter to continue with the untethered flight cards. The pilot retracted the speedbrakes and climbed back to 10,000 ft. MSL.

Untethered Doublets were performed in the both the dirty and clean configurations. The EXD was still two hundred pounds above bingo fuel so the MCC requested the pilot to practice the Bungee Mode Excitation card.

Eclipse
Pilot's Flight Test Report
EXD-01 Flight 5 - First Tethered Flight

At bingo fuel the EXD setup for an uneventful straight-in SFO and full-stop no-chute landing.

Post flight conclusions and recommendations.
Post flight inspection and purging revealed no moisture in the aircraft pneumatic system. Recommend the pneumatic lines from the bottles be vented with dry nitrogen and then closed prior to hookup to the EXD (R13).

Post flight inspection of the release showed the jaws to be nearly closed. Since the release was tested at high loads at angles of up to 20° without any inadvertent releases, the most likely cause of the release was a momentary actuation of the electro-pneumatic switch. Testing of the release under relatively small loads showed that bumping the switch could cause the jaws to open sufficiently for the D-ring to be released without the jaws going over center. The tethered lateral offset required cross controlling of the aircraft and left lateral pressure on the control stick (C14). The pilot flew with his index finger resting against the bottom edge of the release button as an aid in locating the button in the event an immediate release was required. The stability and control of the EXD on tow indicates that an emergency release is extremely unlikely (C15). Recommend the pilot grasp the control stick lower down to preclude the possibility of inadvertent release actuation (R14).

The performance and stability of the tethered aircraft is adequate to ensure safe takeoff from either runway 04 or 22 (C16). Recommend the mission rules be changed to allow takeoffs in either direction (R15). Light turbulence did not pose a hazard to towed operations (C17). Recommend mission rules be changed to allow flight in forcasted and actual light and moderate turbulence conditions (R16). Recommend the mission rules be changed to allow takeoff in winds of up to 20 KIAS with tail or crosswind components not to exceed 10 kts (R17).

The stability and control of the tethered EXD warrants investigation into eliminating the nylon strap center portion of the tow rope (R18).

The on-the-runway portion of the flight operation takes approximately 30 minutes. Mid-day flight operations could reduce the problems that occur due to early morning cold temperatures (C18). Recommend the team pursue clearance for mid-day tethered operations (R19).

Eclipse
Pilot's Flight Test Report
EXD-01 Flight 5 - First Tethered Flight

Conclusions

1. The EXD pilot arrived at the aircraft at the proper time for scheduled pilot entry to find the two chase F-18s were still awaiting fueling.
2. The crew suspected internal icing was inhibiting pneumatic system recharging of the EXD.
3. The EXD pilot felt he was led too far down the runway, and the truck driver felt he should drive referencing the EXD instead of as the "flight lead".
4. The tow rope was not anchored to the base of the pitot boom (as had been done previously), and the crewman handling the rope allowed the knuckle assembly to slam against the stops.
5. Inattention during slack removal could cause high tension values and/or movement of the EXD which could be a hazard to the rope handler.
6. The most likely cause of the nose wheel steering leak was the failure of a seal to initially seat due to the cold weather.
7. The chase's decision to takeoff early avoided any interference with the runway FOD check and allowed time for both he and the video operator to practice positioning.
8. The call for brake release came nearly three minutes after clearance for takeoff, which the EXD pilot felt was excessive.
9. There were no obvious large tension oscillations during the ground roll, and the tow rope remained off the runway at all times.
10. The 170 KIAS takeoff speed was acheived just prior to the C-141A reaching 350 ft. AGL.
11. The canvas sleeving was not required to protect the nylon strapping during ground operations.
12. The EXD was stable in small and moderately banked turns.
13. Separation of the knuckle assembly should be an anticipated consequence of any tow rope release.
14. The tethered lateral offset required cross controlling of the aircraft and left lateral pressure on the control stick.
15. The stability and control of the EXD on tow indicates that an emergency release is extremely unlikely.
16. The performance and stability of the tethered aircraft is adequate to ensure safe takeoff from either runway 04 or 22.
17. Light turbulence did not pose a hazard to towed operations.
18. Mid-day flight operations could reduce the problems that occur due to early morning cold temperatures.

Eclipse
Pilot's Flight Test Report
EXD-01 Flight 5 - First Tethered Flight

Recommendations

1. Recommend additional coordination or emphasis on prompt refueling be completed prior to future early morning operations.
2. Recommend in the future we always have the MC-11 in position to aid in pneumatic charging.
3. Recommend using the procedures used during the CST (truck driver leads the EXD but references the aircraft for his positioning).
4. Recommend the tow rope be anchored to the base of the pitot tube until the EXD is in position and hold.
5. Recommend that once in position and hold, the rope handler remove the anchor at the nose and move to the side of the aircraft where he would be clear of any hazard but still available to assist if required.
6. Recommend future video chase pilots be briefed to takeoff just prior to the tow rope hookup.
7. Recommend the EXD pilot ensures the airborne pickup pilot knows the importance of anticipating the clearance for takeoff and avoiding any unneccesary delays.
8. Recommend the same brake release procedures be used for future takeoffs.
9. Recommend reducing the C-141A rotation and initial pitch attitude to +4° to decrease the altitude at which 170 KIAS is reached.
10. Recommend consideration be made to eliminate the canvas sleeve entirely or limit it to two small sleeves around each of the two-pin connectors.
11. Recommend future test points be completed to investigate the ability for tethered flight at high bank angles.
12. Recommend future operations immediately occur in the PIRA airspace over uninhabited areas.
13. Recommend the pneumatic lines from the bottles be vented with dry nitrogen and then closed prior to hookup to the EXD.
14. Recommend the pilot grasp the control stick lower to preclude the possibility of inadvertent release actuation.
15. Recommend the mission rules be changed to allow takeoffs in either direction.
16. Recommend mission rules be changed to allow flight in forcasted and actual light and moderate turbulence conditions.
17. Recommend the mission rules be changed to allow takeoff in winds of up to 20 KIAS with tail or crosswind components not to exceed 10 kts.
18. The stability and control of the tethered EXD warrants investigation into eliminating the nylon strap center portion of the tow rope.
19. Recommend the team pursue clearance for mid-day tethered operations.

Mark P. Stucky
Eclipse Project Pilot

DAILY/INITIAL FLIGHT TEST REPORT

1. AIRCRAFT TYPE	2. SERIAL NUMBER
C-141A	61-2775

3. CONDITIONS RELATIVE TO TEST

A. PROJECT / MISSION NO	B. FLIGHT NO / DATA POINTS	C. DATE
Eclipse	F-5, 1st Towed Flight	20 DEC 97

D. LEFT SEAT (Front Cockpit)	E. FUEL LOAD	F. JON
Capt Stu Farmer	32,000	C9703900

G. RIGHT SEAT (Rear Cockpit)	H. START UP GR WT / CG	I. WEATHER
Maj Kelly Latimer	200,000 lbs / 28.8%	Clear

J. TO TIME / SORTIE TIME	K. CONFIGURATION / LOADING, SOFTWARE	L. SURFACE CONDITIONS
1622 Z / 0.6 hrs	Petal Doors Removed / Tow Config	37°F/ Winds 230 @ 7kts

M. CHASE ACFT / SERIAL NO	N. CHASE CREW	O. CHASE TO TIME / SORTIE TIME
F-18 NASA 846 / NASA 852		

4. PURPOSE OF FLIGHT / TEST POINTS

The objective of this flight was to perform a towed takeoff and climbout of the C-141 and EXD-01 system. Longitudinal and lateral/directional doublets were performed by the EXD-01 while on tow in climbout and level flight at 10,000 ft MSL.

5. RESULTS OF TESTS (Continue on reverse if needed)

After check in with the NASA 2 control room, the C-141 initiated engine start at 1436 Z. Starting fuel weight was 32,000 lbs. A delay was encountered due to the NASA chase planes requiring fuel service. The C-141 elected to remain engines running and after a short period taxied to a position short of the hammerhead for Runway 04.

At 1526 Z the EXD-01 was approaching engine start and the C-141 GPS data file was reset.

At 1550 Z the C-141 taxied onto the runway to begin the hookup procedure. The rope vehicle arrived at the C-141 at 1601 Z. Hookup was completed and the three pin connector padded. The C-141 was ready to take-up slack at 1607 Z, however, the EXD-01 crewchief noticed a possible nose wheel steering (NWS) hydraulic leak on the EXD-01 when he removed the chocks. This caused a delay while the NWS situation was investigated. At 1615 Z operations continued with the slack removal procedure. The C-141 began its gradual creep forward guided by loadmaster verbal evaluation of the slack. As the loadmaster was counting down toward slack removed, the rope was observed to come up off the ground in tension as the C-141 stopped. Tension was reported by the EXD-01 to be 4000 lbs.

At 1619 Z the tension setting procedure was used to bring tension to 6000 lbs. The NASA control room, the C-141, and the EXD-01 all reported go for flight. Chase established inbound and made 45 and 20 second calls. Countdown to brake release was made and the takeoff roll began at 1622 Z. Takeoff fuel is estimated to be 27,000 lbs. For the desired thrust factor (TF) of 18.0, computed and set EPR was 1.92. Acceleration was brisk. Rotation was to a 5 degree pitch attitude. Calls were made at 200 ft AGL, 170 KIAS, and 350 ft AGL. 350 ft AGL was reached nearly simultaneously with 170 kts. It was later discovered that these calls were heard over C-141 interphone but did not make it out over the radio.

------ CONTINUES NEXT PAGE -------- CONTINUES NEXT PAGE ------

6. RECOMMENDATIONS

C-141 towed takeoff power settings should be confirmed prior to next flight. Either a reduction in power setting, or a shallower climb with a lesser pitch angle should be made to ensure the EXD-01 remains in the ideal takeoff window. On this flight, the C-141 appeared to have a marginally steep climb at takeoff.

The C-141 tail view video camera settings are satisfactory and should remain unchanged for subsequent flights.

A tighter time schedule for day of flight could be developed based on the activity times found during this flight.

COMPLETED BY	SIGNATURE	DATE
Morgan LaVake, Test Conductor		22 DEC 97

AFSC Form 5314 NOV 86 REPLACES AFFTC FORM 385 MAR 84 WHICH WILL BE USED

Document 5. Daily/Initial Flight Test Report, C-141A, 61-2775, 20 Dec. 97, Morgan LaVake, Test Conductor

Section 5 continued.......

During the first couple minutes of the climb, the NASA control room reported dropout of the telemetered data from the EXD-01. This dropout was later attributed to the loss of an electrical generator at the ground reception and retransmission station.

Climbout was made at 190 KIAS. The EXD-01 remained gear down and speed brakes out. During climbout, the EXD-01 performed test cards. The initial card consisted of small control raps in the longitudinal, lateral, and directional axis. No significant results were found. This was followed by longitudinal and lateral/directional doublets in the nominal tow, 4 degrees high, and 4 degrees low positions. Light chop was encountered by the flight passing through approximately 9,000 to 10,000 ft MSL. Prior to completing doublets in the offset to the right tow position, the flight leveled at 10,000 ft MSL.

The C-141 pilot reports no noticeable effects upon the C-141 by the EXD-01 maneuvering.

At 10,000 ft MSL, the doublet cards were repeated in level flight. Fuel weight for the C-141 was 24,000 lbs at 1635 Z. As the EXD-01 was establishing itself for the offset to the right tow position, the aircraft unexpectedly came off tow at approximately 1639 Z. The pilot of the C-141 did feel the "surge" of the EXD-01 tow release. There was initial confusion as to the cause of the release, with the EXD-01 reporting a break of the frangible link, followed by report of tow knuckle separation. Debrief and post-flight review of video footage revealed that that the likely sequence of events was that the electrical release in the EXD-01 was inadvertently actuated. When the knuckle released, the tow rope began violent whipping which caused fracture of the frangible link and loss of the tow knuckle.

The flight then separated with the C-141 continuing to make an uneventful rope drop into the PB8 drop area at 1648 Z.

The C-141 returned to land on Edwards Runway 22 at 1655 Z. Landing fuel weight was 16,000 lbs. The C-141 returned Code 1 with no maintenance writeups.

The test pallet tape was provided to NASA for processing. The NASA Ashtec GPS was removed from the aircraft and returned to NASA for downloading.

Aerodynamics
Eclipse
Flight 5 (1st towed flight)
December 20, 1998

EXD-01 (NASA QEXD-01A 59-0130): Mark "Forger" Stucky
C-141AA (USAF C-141AA 61-7775): Stu Farmer, Kelli Latimer, Morgan
LaVake, John Stahl, Dana Brink, Ken Drucker

The objectives of Flight 5 for aerodynamics were to gather tow
dynamics data. All data was required to operate for mission success.

Aircraft configurations were:

C-141AA: standard; with doors removed, instrumentation pallet
operational, ballast as required, fueled to 225,000 lbs GTOW, and the
low altitude parachute extraction system test pallet for tow
operations.

EXD-01 (QF-106A): standard, with approximately 4000 lbs of fuel,
speedbrake open, and gear down for the entire flight.

Tow train: the towtrain is the nominal configuration with a standard
three pin connector, 475 foot nominal 3/4 inch Vectran rope, two pin
connector, 50 feet of 8 ply nylon, two pin connector, 475 foot nominal
tow rope, and standard end assembly. the two pin connectors are
protected by leather coverings secured and the entire nylon canter
assembly is covered with a large canvas sheath for abrasion
protection.

The flight began with delays waiting for fuel trucks, and a pneumatic
charge system for the EXD-01. Preflight checks were nominal for alpha,
beta, loads r-cal, accel rap test, and knuckle ops check. Chase was
provided by #846 (Ed Schneider/Lori Losey) and 843 (Tom McMurtry).
Weather was cold (28F) and calm; ripples were reported to not be strong
enough as to be characterized as turbulence.

Take off was at 08:26:20, and no calls were heard from the C-141A to assist
the EXD-01 situational awarness; the EXD-01 pilot elected to stay on tow and
continue the mission. The pilot characterized the task as low gain during
the climb. Stick raps and small doublets were performed by the pilot.
Considerable tow rope sail was seen, and the tension was excited throught
the entire flight (possibly due to the canvas sheath flapping?). The MOF
genrator failed during the climb and 26 seconds of data was lost
(08:27:38-08:28:04). Also during the climb, the canvas sheath was seen to be
shredding, and the pilot remained low during the initial stages of the
climb-out to reduce the potential for FOD. The pilot commented that the
aircraft was self recovering, lightly damped, but stable. An inadvertent
release occured at 08:39:10 during a lateral offset test point, and

Document 6. Aerodynamics, Eclipse Flight 5, Al Bowers, Chief Engineer

immediately following the rope began to flail wildly and the knuckle separated on the second oscillation. Post flight analysis indicates this was actuated by the pilot bumping against the release switch on the stick during the cross control required for the lateral offset. The off-tow card 12 was performed and the EXD-01 landed at 08:51:33 after an SFO.

Careful examination of the post flight data showed the aircraft to be markedly more stable than predicted by the simulation, and static trims were seen to be markedly different from the sim as well. The former is believed to be due to mischaracterization of the aerodynamic effects on the rope (the rope sail inducing increased damping). The latter is obviously due to the rope sail. Another adverse effect was the flapping of the canvas sleeve causing the potential for FOD and constant excitation of the bungee mode (this never entirely damped out during the entoire flight, despite the still air).

Take off time: 08 23 38
Release time: 08 39 10
Landing time: 08 51 33
Flight time: 00 27 55
Tow time: 00 15 28
Total Tow Time: 00 15 28

Al Bowers
Chief Engineer

Flight Controls
Eclipse Flight 5
December 20, 1998 7

Brake release and initial roll.
After brake release by the C-141A, the pilot of the EXD-01 airplane smoothly
released his brakes from a pre-load tension of 6,000 lbs. Initial longitudinal
acceleration was less than .2 g s. The longitudinal bungee mode was
immediately excited at a frequency of approximately 1.6 radians/sec. Maximum
tow rope tension at the second peak of the bungee oscillation was slightly less than
13,000 lbs. Subsequently, the bungee mode damped out in about five cycles. At the
same time the tow rope tension gradually decreased to less than 7,000 lbs. After
an initial roll lasting for about seventeen seconds, the EXD-01 entered the wake of
the C-141A. The most pronounced effect of the wake was on the angle of attack
and sideslip vanes, resulting in large amplitude oscillations (−10 degrees), but no
significant rigid-body angular velocities since the EXD-01 was still on the ground.

EXD-01 rotation and takeoff
The rotation of the EXD-01 airplane was preceded by the takeoff of the C-141A at
approximately 115 knots. Rotation of the EXD-01 was initiated while still in the
wake of the C-141A; however, when the latter was reached an altitude of
approximately 200 feet, the EXD-01, still on the ground, was completely clear of the
wake. The EXD-01 became smoothly airborne at 170 knots. Neither the 200 feet
nor the 170 knots calls by the C-141A were audible in the control room. During
rotation and takeoff of the EXD-01 the tow rope tension increased gradually from
7,000 to 12,000 lbs. Thereafter the tow rope tension decreased and remained below
10,000 lbs for the rest of the flight. Immediately after takeoff by the EXD-01
airplane the indicated tow rope elevation angle reached a high value of about 13
degrees. This is higher than the steady state trim values predicted by the
simulator. Elevation angles greater than 10 degrees required positive, i.e. nose
down, elevator angles. Again, positive elevator angles during takeoffs had not been
observed in the simulator. During takeoff the altitude difference between the C-
141A and the EXD-01 reached 400 feet, the largest value during the entire towed
portion of the flight.

Climb and test points
During stabilized towed flight it was noted that the tow rope elevation angle was
considerably larger than the simulator prediction, requiring more nose down
elevator. Post-flight video playback of a camera located in the C-141A showed that
the tow rope in flight had the shape of a convex catenary. This shape would
account for the higher tow rope elevation angles. The tow rope tension during
stabilized level flight and mild maneuvering flight was approximately 30 percent
lower than simulator predictions. Whether this is caused by higher idle engine
thrust or lower in-flight drag requires further analysis. During the climb the
average value of the vertical separation of the two airplanes was approximately
280 feet; the +4 and -4 degree tow rope elevation increments from the nominal
value corresponded to approximate altitude differences of 310 feet and 200 feet,
respectively. The altitude differences that are being quoted here refer to the GPS

Document 7. Flight Controls, Eclipse Flight 5, Joe Gera

antenna positions rather than the c.g. positions of the two airplanes. Other than the trim elevator and tow rope elevation angle differences, the pitch and lateral-directional doublets were similar to simulator predictions, especially up to the point where the protective canvas cover of the nylon webbing started to shred off the tow rope. Afterwards, it appeared that the low-amplitude bungee oscillation was set off at times without any pilot input. It was noted that the bungee oscillation occurred mainly in the pitch plane of symmetry. A possible cause of this may be the wider frequency separation of the dutch roll mode ($w_n \sim 1.9$ rad/sec) from the bungee frequency ($w_n \sim 1.45$ rad/sec) than that of the short-period mode ($w_n \sim 1.7$ rad/sec).

Tow rope release
An unexpected release occurred during test point 5, card 7. Tow rope tension at release was approximately 6,300 lbs. No undesirable transients were noted by the pilots of either the C-141A or EXD-01 during the release; however, all of the metal hardware was lost from the end of the tow rope during the subsequent large-amplitude oscillations of the rope.

Untethered test points.
After tow rope release longitudinal and lateral-directional doublets were performed by the EXD-01 at the desired test conditions in the landing and the clean configurations.

Comments
This was a highly successful first towed flight of a pure delta-wing airplane of relatively high wing loading. Because of the simulation and the high-speed taxi test, there were few surprises in the stability and control area. Differences between simulation and flight can be attributed principally to the fact that the actual shape of the tow rope was other than a straight line along the line-of-sight between the attach points. This was accentuated be the presence of the 50-foot long canvas shroud in the middle of the tow rope as indicated by in-flight photographs of the tow rope.

Joe Gera

Eclipse
Pilot's Flight Test Report
EXD-01 Flight 6 - Second Tethered Flight, 21 Jan 1998

Overview
This flight was a near repeat of the first tethered flight, the major change being the streamlining of the canvass sleeving which covered the 50 ft. section of 8-ply nylon strapping in the middle of the tow train. The new canvass sleeve was necked down to more tightly fit the strapping and did not cover the rear 2-pin connector (or extend aft beyond it). This change decreased the drag of the center section, eliminated any flapping of the canvass, and virtually stopped the tow rope "quiver."

Regardless of maneuvers the slip indicator normally showed a 1/4 to 1/2 left ball deflection. I tried trimming it to center, but it seemed unnatural so I left it as it was.

Ground Operations
The ground operations showed improvement over Flight 5 and went like clockwork once the aircraft took the runway. The EXD had initially been fueled to 5000 lb., but an extra 1000 lb. was added to allow us more time for research test points. I reported a 1500 lb. fuel imbalance (left side 2700 lb., right side 1200 lb.) on power-up. The test controller reported the control room didn't have any issues with this, which I interpreted to mean that we would not take any corrective measures and see what happened over time.

Tethered Takeoff
The established chase procedures worked well, and the brake release call was expeditious. The initial takeoff roll tension was controlled, but not as smoothly as on Flight 5.

Takeoff acceleration seemed slower than during the taxi test and first tethered flight, particularly the portion after EXD rotation. Takeoff occurred approximately 1000 ft past the midfield taxiway despite the absence of a tailwind. Post-flight debrief with the C-141A crew failed to illuminate any reason for this apparent decrease in performance. Capt. Farmer used the recommended procedural change (after Flight 5) of decreasing the pitch attitude 1° after raising the gear and flaps to facilitate reaching 170 KIAS at a lower AGL altitude.

The C-141A radio calls during the takeoff matched well with my visual estimations of declination. The heavier left wing was felt and required four clicks of lateral trim to balance the forces. In retrospect, we should have **attempted to balance the fuel prior to flight.**

The tow rope appeared significantly quieter WITH the new canvass sleeved configuration. There was less apparent rope sail and no quivering of the nylon center section.

Flight Cards
Climb Rate Effects on Stability. Initially ARRIS reported setting 1000 fpm climb but it settled out at 1500 fpm, so the card was done as planned. I estimated a -16° declination at 1500 fpm. The tow rope seemed very stable as

Document 8. Eclipse Pilot's Flight Test Report, EXD-01 Flight 6—Second Tethered Flight, 21 Jan. 1998, Mark P. Stucky, Eclipse Project Pilot

Eclipse
Pilot's Flight Test Report
EXD-01 Flight 6 - Second Tethered Flight, 21 Jan 1998

compared to the first tethered flight. When the climb rate was decreased to 1000 fpm the trim position seemed to decrease to an estimated -14°. Reestablishing the same declination required a slight nose down trim. The Loadmaster observed the rope to be lightly resting on the skid plate and not "flying" like it had on the first flight. Total fuel at completion was 4500 lb., and lateral asymmetry was reduced to 1200 lb..

Stability Boundary Investigation (dirty configuration). I was able to descend to the lower gimbal angles with the only handling qualities difference being a slight lateral "looseness". Climbing upward the rope appeared straight at an estimated -5° declination, and the Loadmaster reported it to be one foot above the ramp. The test controller reported the "gimbal angles," and I mistook the call to mean that I had reached the upper gimbal angle so I aborted the point. It was only on a later card that I realized the upper gimbal limits had not been reached. This is a reminder that we should never use a term which is a "directive" call as part of "descriptive" call.

Drag Effects on Stability. As noted on 5, the EXD airspeed indicator indicated 10 kts less than the C-141A's sensitive indicator. Retracting the speedbrakes did cause a cyclical tension oscillation as well as a slight downward pitching moment (2 clicks of nose up trim required). Opening them resulted in a 1500 lb. tension increase. The aircraft was significantly quieter and airframe noise eliminated when the landing gear were retracted. This was apparent in the control room's stripchart data as well. The drag difference associated with the landing gear seemed on par with that of the speed brakes. When the gear and speed brakes were retracted, the white light illuminated with 2000 lb. of tension. The tension stabilized at 3500k lb. in the clean configuration. At the completion of this card the fuel imbalance had decreased to 1000 lb.

HQR Criteria - On Tow (dirty configuration). The C-141A reported a 1/4 ball deflection when I lined up behind the number 4 engine. The HQR task was easy to accomplish. I was able to be moderately aggressive in the initial bank angle to capture the centerline. There was also a natural centering effect due to the tether. It was relatively easy to capture centerline with small or no overshoots and HQR ratings of 2. I did feel that if I was overly aggressive it could cause me to get out of phase with the system dynamics and perhaps enter an aircraft pilot coupling.

Trim in Turns (dirty configuration). The EXD was very stable in turns of up to 40° angle of bank (hands off stability). At 45° I felt the EXD was becoming spirally unstable. This may be true, or it could have been caused by being slightly out of position (high / outside of turn). Being slightly out of position is not easily apparent; the best way to verify it would have been to relax the lateral input and see if the EXD would stabilize at the new position. Unfortunately, we rolled out of the turn prior to attempting this. In subsequent turns we never got above 40° bank, so I could not reevaluate the spiral stability at 45° again.

Eclipse
Pilot's Flight Test Report
EXD-01 Flight 6 - Second Tethered Flight, 21 Jan 1998

Upper Tow Limit / Wake Investigation (dirty configuration). I was confused over the point of starting this card, becasue I mistakenly felt we had reached upper gimbals during the stability boundary investigation card. Once the situation was clarified, I started climbing. I offset to the right for increased clearance between the pitot tube and tow rope. I was able to climb into the wake where the Loadmaster reported the rope appeared to be straight out the back of the C-141A. Chase reported the rope looked very close to the pitot boom (the lateral offset was not apparent from his side view). Flight in the wake appeared to be slightly easier than in the off-tow condition, as there was no uncommanded yaw/sideslip. Small lateral inputs were required to counter uncommanded rolling motions.

Lateral Offset (dirty configuration). Maintaining the lateral offset required cross controlling and doing the doublets from this offset "neutral" condition.

Stability in Descent (dirty configuration). At -500 fpm a slight pitch oscillation was noted in the upper (-5° declination) position. It was still possible to trim for hands off flight. The rope sail appeared greater at the 1000 fpm descent nominal position. In the lower position a large amount of rope sail existed and a small approximately 3 Hz lateral oscillation was observed.

We set up to return to card 3 when ARRIS surprised us with a call indicating they needed to RTB in ten minutes. We therefore proceeded with the tow release. **To facilitate research efficiency I recommend the C-141A crew provide joker & bingo values in the flight crew brief and include them in the flight fuel checks.**

Normal Tow Release. This card was done at a slower airspeed (170 KIAS) and lower altitude (2,500 ft AGL). The reasoning was the slower airspeed may decrease the rope whipping energy and increase the chance of the knuckle remaining on the tow rope. The lower altitude was the planned C-141A rope drop altitude and would decrease the hazard area as well. The NASA team had hoped the tow release could be immediately followed by the rope drop, but the C-141A crew declined in the crew brief indicating they needed additional time to setup.
    The transition to 170 KIAS was easy to accomplish, and no transients were noted when the C-141A lowered their flaps. The cockpit AOA gauge indicated 1/3 of the way between final approach and minimum safe speed. An offset to the right was done, and the electro-pneumatic tow release actuated. The release was clean at approximately 5000 lb. of tension. It was easy to clear the rebound of the rope. The knuckle did stay attached, and the whipping seemed of a lesser intensity. The C-141A turned to enter the downwind leg for their rope drop, but unfortunately the knuckle whipped free of the tow rope prior to the planned drop. **Recommend the C-141A crew try to accommodate NASA's request to meld the tow release with their rope drop procedure. Recommend we consider intentional breakage of the frangible link as a method to retain the instrumented knuckle.**

Eclipse
Pilot's Flight Test Report
EXD-01 Flight 6 - Second Tethered Flight, 21 Jan 1998

Air Data Calibration.   Accomplished as planned.

RTB and landing was uneventful.

Summary of Lessons Learned and Recommendations
1. We should have attempted to balance the fuel prior to flight.
2. We should never use a term which is a "directive" call as part of "descriptive" call.
3. To facilitate research efficiency I recommend the C-141A crew provide joker & bingo values in the flight crew brief and include them in the flight fuel checks.
4. Recommend the C-141A crew try to accommodate NASA's request to meld the tow release with their rope drop procedure.
5. Recommend we consider intentional breakage of the frangible link as a method to retain the instrumented knuckle.

Mark P. Stucky
Eclipse Project Pilot

DAILY/INITIAL FLIGHT TEST REPORT		1. AIRCRAFT TYPE	2. SERIAL NUMBER
		C-141A	61-2775

3. CONDITIONS RELATIVE TO TEST

A. PROJECT / MISSION NO	B. FLIGHT NO / DATA POINTS	C. DATE
Eclipse	F-6, 2nd Towed Flight	21 JAN 98
D. LEFT SEAT *(Front Cockpit)*	**E. FUEL LOAD**	**F. JON**
Capt Stu Farmer	34,000	C9703900
G. RIGHT SEAT *(Rear Cockpit)*	**H. START UP GR WT / CG**	**I. WEATHER**
Maj Kelly Latimer	202,000 lbs / 28.8%	Clear
J. TO TIME / SORTIE TIME	**K. CONFIGURATION / LOADING, SOFTWARE**	**L. SURFACE CONDITIONS**
1531 Z / 1.2 hrs	Petal Doors Removed / Tow Config	27°F/ Winds 260 @ 6kts
M. CHASE ACFT / SERIAL NO	**N. CHASE CREW**	**O. CHASE TO TIME / SORTIE TIME**
F-18 NASA 846		

4. PURPOSE OF FLIGHT / TEST POINTS

This flight was the second towed flight of the Eclipse program. Configuration differed from the 1st flight in that the center nylon webbing section of the tow rope was more tightly sheathed in its cover in an effort to reduce the aerodynamic turbulence, instability, and rope sail seen in the rope on the first flight. Tethered test points explored climb rate, descent, turns, the stable tow envelope in both clean and dirty configurations, lateral offset, and the effects of varying the EXD-01 drag (by configuration changes).

5. RESULTS OF TESTS *(Continue on reverse if needed)*

The C-141 experienced a 15 minute delay in engine start due to an apparent problem with the pilot's attitude indicator. Maintenance believes the problem may have been caused by the tie in of the test pallet instrumentation to the system. It is reported to historically have problems operating in cold conditions.

The C-141 had an uneventful start. At 1450 Z the C-141 began taxi. Shortly after the C-141 reached the RWY 04 hammerhead, the EXD-01 was nearing ready to taxi and the C-141 took the runway at 1502 Z. Hook-up proceeded smoothly and the rope truck provided the connector to the C-141 at 1511 Z. At 1517 Z hookup of the connector on the C-141 was complete and the C-141 was ready to take up slack. As the loadmaster was counting down over interphone for the slack removal, the EXD-01 called for the C-141 to stop. This stop occurred slightly prior to total slack removal. Post flight debrief determined that the loadmaster will in the future make his call on the radio rather than interphone.

The tension setting procedure was used to bring tension to 6000 lbs. The NASA control room, the C-141, and the EXD-01 all reported go for flight. Chase established inbound and made 45 and 20 second calls. Countdown to brake release was made and the takeoff roll began at 1530 Z. Takeoff fuel is estimated to be 31,000 lbs. For the desired thrust factor (TF) of 18.0, computed and set EPR was 1.92. Take-off was uneventful.

Climbout was at 190 KIAS. The power was reduced to idle on the outboard engines at 185 KCAS to remain below 200 KCAS in the climb. Airspeed was maintained primarily by small adjustments in pitch attitude. The climb rate effects test points (trim shots) were taken at 1500 fpm and 1000 fpm rates of climb. The flight leveled at 10,000' MSL.

------ CONTINUES NEXT PAGE ------

6. RECOMMENDATIONS

Pre-heat of the cargo compartment and test pallet should be considered in an effort to increase the reliability of the test instrumentation.

Based on the knuckle assembly remaining on the rope for well over one minute after release of the EXD-01, it is likely possible to procedurally set up for and expedite drop of the rope over the drop zone prior to knuckle separation, thus retaining the knuckle hardware.

COMPLETED BY	SIGNATURE	DATE
Morgan LaVake, Test Conductor		22 JAN 98

AFSC Form 5314 NOV 86 REPLACES AFFTC FORM 365 MAR 84 WHICH WILL BE USED

Document 9. Daily/Initial Flight Test Report, C-141A, 61-2775, 21 Jan. 98, Morgan LaVake, Test Conductor

Section 5 continued.......

Throughout the tow operations, the pilot commented on a low rumbling noise present that appears to be transferred into the C-141 airframe through the rope. No noticable pitch motions or longitudinal accelerations were present until the release.

The flight proceeded with the stability boundary investigation and drag effects on stability cards. Tension in the clean configuration was reported to be approximately 3000 lbs. The turn card was completed at bank angles of 15, 30 and 45 degrees.

During the HQR Criteria - On Tow card, the C-141 pilot reported a slightly noticeable slip with about one quarter ball deflection of the slip indicator when the EXD-01 was offset behind an outboard engine.

At 1558 Z C-141 fuel load was 21.8 Klbs. The upper tow limit and lateral offset cards were then completed. At 1605 Z the C-141 began a decent to 5000' MSL and the descent card was completed in the decent at rates of 500 fpm and 1000 fpm. The rope sail was reported noticeably increased during the descent. The loadmaster reported the rope to be 1 to 1.5 ft above the ramp edge in the decent with the EXD-01 in the 4 degrees higher than nominal tow position.

At 1610 Z a climb was initiated to repeat the drag effects card. However, C-141 fuel at this point was 18.8 Klbs and it was decided to proceed with the mission with no further delay or repeat.

The flight descended to 5000' MSL and slowed to 170 KIAS for the release. At 1615:50 Z the EXD-01 was released. The release was characterized as benign with the longitudinal acceleration defined as "gentle push". The rope generated a large amount of slack with the kickback but no portion of the rope came in contact with the C-141 aircraft. After several seconds, the rope quickly settled down trailing behind the aircraft. The first half of the rope was smooth until reaching the webbing with motions gradually increasing until reaching the rope end which was whipping violently.

At 1618 Z the scanner reported that the knuckle had just separated from the end of the rope. Approximate coordinates were 34°53.5, 117°43.7. Winds as reported by chase were 020 deg at 12 kts. The C-141 continued with the rope drop, dropping it onto the center of PB8 drop zone at 1622 Z.

The C-141 rejoined with the EXD-01 for the HQR Criteria - Off Tow card. The chase radar was reported to have stopped operating and the EXD-01 established itself at the 1000 ft aft point by visual reference. Time was 1627 Z and C-141 fuel was at 15.6 Klbs. Test cards were completed at 1629 Z and the formation separated for RTB.

The C-141 returned to land on Edwards Runway 22 at 1638 Z. The C-141 returned Code 1 with no maintenance writeups.

The test pallet tape was provided to NASA for processing. The NASA Ashtec GPS was removed from the aircraft and returned to NASA for downloading. Postflight processing of the test pallet tape revealed it to be blank with the exception of a two second segment at the beginning. It is surmised that this brief recording on it is part of the preflight when the tape was loaded. The blank tape cannot be properly explained. The pallet operator confirms that the tape was started and the record light was on. Instrumentation technicians inspected the pallet and made a test tape which was found to be satisfactory.

Aerodynamics
Eclipse
Flight 6 (second towed flight)
January 21, 1998

EXD-01 (NASA QEXD-01A 59-0130): Stucky
C-141A (USAF C-141AA 61-7775): Farmer, Latimer, LaVake, Stahl, Brink, Drucker

Objectives of this test were reprioritized from the first test flight. In the first test mission objectives were, 1) dynamic response, 2) handling qualities, and 3) static trim. Based on results of the first test mission, the prioority was now on static trim, dynamics response and handling qualties. The potential for adding characterization of the rope sail/catenary was also examined, though no action was yet taken for flight.

Aircraft configuration was the same as for Flight 5, with the following exceptions:

C-141A: no change.

EXD-01: no change, except for selected points which would be made with the aircraft clean (speedbrake retracted and gear up).

Tow train: the canvas sleeve was fabricated much smaller and tighter around the nylon straps; reinforcements were added to prevent failure of the sleeve in flight. The ends of the sleeve were secured between the leather protection for the two pin connectors.

No unforseen delays happened during ground ops. Chase was provided by #846 (Ed Schneider/Lori Losey). There was no perceptible turbulence. The day was cold (about 30F) and calm. Takeoff of the EXD-01 was at 07:31:13 hours. Trim shots on climb were made at 1500 fpm and 1000 fpm rate of climb. Nominal test altitde remained 10k ft msl and 190 KEAS, and air space over the PIRA was used as much as reasonably possible. During the climb, it was obvious the rope was not sailing as much as before and was much closer to a straight line between the C-141A and the EXD-01. Some lateral looseness was experinced by the pilot concurrent with the contact of the wake behind the C-141A. Trim in turning flight was performed, the spiral mode was noted to become unstable at 45 degrees angle of bank, but not at 15 or 30 degrees. The aircraft was cleaned up and transient tension values as low as about 2200 lbs were noted.

An HQR task of lateral offsets were performed twice, Cooper-Harper Ratings of 2 were given. A climb through the wake turbulence was performed. A lateral offset card was performed. Trim shots in stabilized descent was performed at 500 fpm and 1000 fpm. Some lateral oscillations were noted at the 1000 fpm point. The rope was released at 08:16:50. Rope flail was again observed, though a lower airspeed release was used, and the knuckle did not depart the end of the rope behind the C-141A until 08:18:30. An off tow HQR task was repeated, and ratings of 4 and 2 were given. The EXD-01 landed at 08:34:32.

Take off time: 07 31 13
Release time: 08 16 50
Landing time: 08 34 32
Flight time: 01 03 19
Tow time: 00 45 37
Total Tow Time: 01 01 05

Al Bowers
Chief Engineer

Flight Controls
Eclipse
EXD-01 Flight #6
January 21, 1998

Brake release and takeoff.
Initial longitudinal acceleration was smooth, approximately .2 g s. The tow plane wake was encountered during the takeoff roll, just like in the previous flight. By the time the EXD-01 reached a speed of 150 knots, the wake had already passed overhead so that the rotation and takeoff took place in smooth air.

Climb to 10,000 feet.
The climb was made at 1,500 and 1,000 ft/min climb rates. The only effect of the climb rates was a slight increase in tow rope tension during the climb in comparison with level flight.

Stability boundary tests.
No instability was encountered in either the high- or low tow positions. At the low position the 20-degree gimbal limit was reached. The pilot also reported a lateral looseness while in the lower than nominal tow position.

While the stability boundaries were probed, video coverage of the tow rope showed that the rope can assume the shape of a convex or a concave catenary, or even that of a straight line, depending on the vertical separation of the two airplanes.

Drag cleanup tests.
From the nominal gear-down and speedbrakes-out configuration, transition was made incrementally to the clean configuration. Approximately equal tow rope tension reductions were noted with the retraction of the speed brakes and the landing gear. Both increments were comparable with simulator predictions; however, each value of the tow rope tension was approximately 1,800 lbs less than the corresponding simulator prediction.

In the clean configuration the amplitude of the bungee oscillation was significantly reduced to the point that it was hardly noticeable in the tow rope tension time history. The lowest value of the tow rope tension, approximately 2,500 lbs, occurred in the clean configuration.

No degradation in controllability or in the handling qualities of the airplane was noted in any of the reduced-drag configurations.

Turning flight.
At various times throughout the flight turns at bank angles of 15, 30, and 45 degrees were made by the C-141A tow plane. The EXD-01 had no difficulties while following the tow plane in turns. No significant trim changes or sideslip angles were noted any time in turning flight.

Document 11. Flight Controls, Eclipse Flight 6, Joe Gera

Descents on tow

Rates of descent of 500 and 1,000 ft/minute were set up by the tow plane. There was no tendency for tow rope slack during these descents, although there was some lateral oscillations of the tow rope noted by the EXD-01 pilot during a downward excursion from the nominal tow position.

Handling qualities tests.

From lateral offsets, both on and off tow, the pilot aggressively flew back to the center to capture the nominal tow position. The pilot rated all of his maneuvers a 2 on the Cooper-Harper scale. He noted that the task while on tow was considerably different because he only had to relax the stick for the tow rope to pull him back to the center. Off tow aggressive roll stick inputs were needed to reacquire the nominal tow position.

Tow rope release.

In an attempt to impart less energy to the tow rope at release, the towing airspeed was reduced by the C-141A by approximately 20 knots prior to release. The tow rope release hardware however stayed on the tow rope for only about a minute and a half during the violent oscillations subsequent to the release.

Joe Gera

Structures Report
EXD-01 Flight 6
January 21, 1998

This flight was structurally mild. Drag effects on EXD-01 stability were studied. Load data quality was good and the zero load offsets for the prime and spare channels were about +100 and -100 pounds respectively. This was a smaller offset (better) than was produced during the first tethered flight. Tow rope lift off load during tensioning procedure was about 5,500 pounds - similar to first tethered flight.

System Configuration was similar to the previous flight except for a revised center nylon segment sheath. The new sheath was made of a higher strength material and was fitted snugly so that it did not flap in the air flow. This new design appeared to reduce rope sail and the aft edge of this sheath did not shred as on the previous flight. The sheath is not primary load bearing structure, but serves to protect the nylon strap assembly from possible abrasion during the take-off roll and from flapping in the wind or buzz during up and away flight. This new design was a noticeable improvement.

The highest tow load of the flight was about 13,800 pounds which occurred during the early part of the take-off roll on the third cycle peak which was about ten seconds after EXD-01 brake release. No slack was observed.

A significant finding of this flight was that going from dirty to clean EXD-01 configuration not only reduced the tow load, but improved stability (as shown by the much smoother load signal) as well. This was not what was expected. A noticeably stable tow rope load of about 3,500 pounds, the lowest of this flight, was produced with the landing gear up and the speed brakes in.

A commanded normal release of the tow rope at the EXD-01 was accomplished using the electro-pneumatic system. Release was performed at a tow rope load of about 6,000 pounds and with a lateral offset. Rope separation was positive and adequate. After some rope whipping the end assembly departed from the rope after breaking the frangible link (apparently in low cycle bending). The end fitting did not tear through the rope loop and was recovered with the rope. The separated end assembly was not recovered.

Bill Lokos
Structures Engineer

Document 12. Structures Report, EXD-01 Flight 6, 21 January 1998, Bill Lokos, Structures Engineer

Weather Summary
Eclipse
EXD-01 Flight #6
January 21, 1998

Building surface and upper-level high pressure with a weak offshore flow over Southern California was the situation leading to the 2nd ECLIPSE tow flight on Wednesday, January 21, 1998. Due to a passage of a cold front on Monday, the air was cold and stable resulting in chilly overnight temperatures and some hazy skies. This cold and stable weather was forecast to persist on flight day with variable winds with low or no turbulence. Sky coverage was forecast to be scattered to broken at 20,000 feet. Overall conditions were expected to be good during the flight

Flight day observations were very close to forecast. The minimum temperature observed by the runway wind towers ranged from 27 deg F to 32 deg F. The wind speeds peaked at a value of 9 knots just before 0200 PST then gradually decreased to 3 knots by 0600 PST. The wind speeds at takeoff time, 0731 PST, were observed to be 2 to 3 knots out of the west of southwest. The surface winds did not exceed 4 knots throughout the flight. The wind and temperature data were observed from wind towers located along the main runway. Tower 044 is located 4000 feet down the threshold of runway 04 and Tower 224 is 4000 feet down runway 22. Both towers measure data 30 feet above ground level. No turbulence was reported during the flight. Cloud coverage was observed as broken at 25,000 feet. No severe weather or precipitation was observed in the area.

Casey Donohue

Document 13. Weather Summary, Eclipse, EXD-01 Flight 6, 21 January 1998, Casey Donohue

Eclipse
Pilot's Flight Test Report
EXD-01 Flight 7 - Third Tethered Flight, January 23, 1998

Overview
This flight was similar to Flight 6, the major change being the elimination of the 50 ft. section of 8-ply nylon strapping in the middle of the tow train. The all-Vectran tow train had decreased damping characteristics, which caused more tension spikes and increased workload during the takeoff roll. Flying qualities still remained very good to outstanding while within the stable region of low tow.

Ground Operations
Once again we had a fuel imbalance, however this time we shutoff the low side boost pumps and were able to have it nearly balanced prior to takeoff. Ground operations went well and the MC-11 was used for starting. It was obvious during the slack removal and tensioning that the all-Vectran rope had much less shock absorption than the previously used nylon strapping and Vectran combination. The transition from an unloaded and slack rope to a loaded taut rope was observed to be more abrupt by both the Loadmaster and myself. We were ready for takeoff at 0721. The center section of the rope came off the ground at 3000 lbs. of tension (as predicted).

Tethered Takeoff
The tension spikes during the initial ground roll were more abrupt but were well controllable by close attention to the tension gauge. Once the tension was stable during the roll, the rest of the takeoff felt very similar to previous tethered takeoffs.

Takeoff acceleration seemed to match the slower takeoff profile seen on the previous flight. Takeoff temperature was 32° F.

During climbout I noticed a small (approximately 1/2°) continuous pitch oscillation.

The tow rope visually appeared very similar to the Flight 6 configuration (tailored canvas sleeve). That is to say that while there was a very noticeable difference in rope sail and rope "noise" between the large canvas sleeved configuration and the tailored canvas sleeve, there was much less of a visual difference (if any) between the latter and the all-Vectran configuration.

Flight Cards
Climb Rate Effects on Stability. No new comments.

Stability Boundary Investigation (dirty configuration). After level off, I was at an approximately 10° declination and the EXD exhibited very stable hands free stability. I descended to the lower limit (gimbals) and climbed slightly in order to do the doublet series. Approximately 20 feet of rope sail was apparent. The pitch response was lightly damped and slowly coupled into an unstable lateral/directional oscillation. The lat/dir response was a quicker entry into

Document 14. Eclipse Pilot's Flight Test Report, EXD-01 Flight 7—Third Tethered Flight, 23 January 1998, Mark P. Stucky, Eclipse Project Pilot

Eclipse
Pilot's Flight Test Report
EXD-01 Flight 7 - Third Tethered Flight, January 23, 1998

the unstable lat/dir mode. It was easy to immediately stop the oscillation with pilot input.

I climbed up to the medium high (straight rope) position. The Loadmaster reported the rope was one foot above the ramp. Response to doublets was stable in all axis.

In the upper position (just below the wake) the bungee mode was excited and resulted in pitch oscillations. These oscillations were dramatic in the EXD and could be felt in the C-141A. The response felt similar to that experienced in the simulator. I did not feel it was prudent to do a doublet but was able to go stick fixed for several cycles. The period appeared to be approximately four seconds long. The Loadmaster reported the rope to be 1.5 to 2 feet above the ramp with a lot of rope motion on his end.

Drag Effects on Stability. Similar effects as seen on Flight 6 with perhaps slightly sharper initial oscillations and longer damping.

Stability Boundary Investigation - Clean. More rope sail was observed in this lower drag configuration. At the lower limit the rope the Loadmaster reported the rope was lightly resting on the ramp. I estimated 30 feet of rope sail. I felt a bungee-type oscillation which was not felt in the C-141A. I also commented that the stability seemed better than in the dirty configuration.

Next I climbed upward to the medium high (straight rope) position. The Loadmaster reported the rope to be 8" above the ramp. The aircraft had a stable response to the doublets but an unstable lateral motion caused by a slight heading differential caused by the doublet response. This motion had a much longer period than a pure aircraft response. The EXD's heading would diverge slightly until the rope pulled tight and jerked the nose around. Then the EXD would reverse direction and the process would repeat with increasing severity.

In the high position (just below wake) the bungee mode was easily excited and the EXD's handling qualities felt looser. The Loadmaster reported the rope to be one foot above the ramp.

HQR Criteria - On Tow. Both clean and dirty tasks were performed. Both were easy tasks, which I gave ratings of "2" to. Despite the equal ratings, I would rate the clean task between the dirty task and off-tow task. There is less restoring force as the tow tension decreases which makes for a longer, more tame capture.

Trim in Turn Effects. These turns were all done in the near-nominal position. A small lat/dir then pitch oscillation was felt in the first 15° banked turn. Increasing the turn to 30° seemed more stable with only a 1-2° lateral oscillation and +/- 500 lbs. tension oscillations. The 45° banked turn was actually closer to 40° and was still spirally stable.

Eclipse
Pilot's Flight Test Report
EXD-01 Flight 7 - Third Tethered Flight, January 23, 1998

<u>Upper Tow Limit / Wake Investigation - Dirty Configuration</u>. As in Flight 6, I offset laterally for clearance between the pitot boom and rope. This time I climbed higher. It took nearly full aft stick to ascend into the upper reaches of the wake. There was insufficient remaining control authority to climb above the wake or counter the roll effects of the vortices. I had two vortex encounters, both of which dropped the EXD back below the wake. At the upper position the Loadmaster reported the rope to be 2.5 - 3 ft. above the ramp.

<u>Stability in Descent</u>. No additional comments from Flight 6.

<u>Frangible Link Separation Maneuver</u>. Prior to attempting release, we did several build-up maneuvers. The first time I added power and moved forward slightly a very large amount of sail occurred. The rope was stable, and I was confident I could generate much more slack without any tendency for the slack to be a hazard to the EXD. It appeared the aerodynamic forces on the rope caused it to bow upward in the center section. This was easy to mistake for slack, but when I reduced power and the rope tightened the peak load was only 8000 lbs.

The next attempt was taken further with a 13,000 lbs. tension spike. The third attempt reached 21,500 lbs. It was easy to maintain control during the tension transients. With these amounts of slack the knuckle is at its gimbal limits, but the tension is negligible.

Next, we set up for the planned release. The release occurred with a loud bang, just as occurs with a normal release. I cleared to the right. The knuckle assembly remained stable with the remaining portion of the frangible link remaining cocked up under the aerodynamic loads.

RTB and landing was uneventful.

<u>Closing Comments</u>. This was another very successful flight. Procedures worked well, and I have no additional recommendations.

Mark P. Stucky
Eclipse Project Pilot

Eclipse
EXD-01 Flight 7
Third Tethered/Release Flight
January 23, 1998

Project Manager's Comments

There will be very few days like this one for me in my career in flight research. This mission was perfectly executed. And the data we obtained was extremely interesting.

Throughout this project I can honestly say that I think we have operated as a very effective and integrated team. But there are occasions when individuals and their performance stands out. When I reflect on today's effort, it is my assessment that our pilot, Mark Stucky (Forger), was the kingpin. For the past missions, we have all performed our various tasks and responsibilities. Yet today, it was very clear to me that Forger's expertise and judgment made it happen. We obtained all of our test points. The points were executed flawlessly. We used a new rope configuration. The frangible link was broken without a problem of any kind. This was a perfect mission. Forger, I want to offer you my sincere compliments. And I know that I speak for the Project when I say this.

I am glad that you all were able to participate in this rare research event.

Carol A. Reukauf

Document 15. Project Manager's Comments, Eclipse, EXD-01 Flight 7, Third Tethered/ Release Flight, 23 January 1998, Carol A. Reukauf

DAILY/INITIAL FLIGHT TEST REPORT

1. AIRCRAFT TYPE	2. SERIAL NUMBER
C-141A	61-2775

3. CONDITIONS RELATIVE TO TEST

A. PROJECT / MISSION NO	B. FLIGHT NO / DATA POINTS	C. DATE
Eclipse	F-7, 3rd Towed Flight	23 JAN 98

D. LEFT SEAT *(Front Cockpit)*	E. FUEL LOAD	F. JON
Capt Stu Farmer	35,000	C9703900

G. RIGHT SEAT *(Rear Cockpit)*	H. START UP GR WT / CG	I. WEATHER
Maj Kelly Latimer	203,000 lbs / 28.8%	Clear

J. TO TIME / SORTIE TIME	K. CONFIGURATION / LOADING, SOFTWARE	L. SURFACE CONDITIONS
1523 Z / 1.4 hrs	Petal Doors Removed / Tow Config	29°F/ Winds 020 @ 3kts

M. CHASE ACFT / SERIAL NO	N. CHASE CREW	O. CHASE TO TIME / SORTIE TIME
F-18 NASA 846 & NASA 852		

4. PURPOSE OF FLIGHT / TEST POINTS

This flight was the third towed flight of the Eclipse program. Configuration differed from the 2nd flight in that the nylon webbing damper section of the tow rope was removed. The tow rope on this flight was a continuous Vectran rope. Also differing on this flight was the separation. Rather than a normal release of the knuckle assembly from the F-106, the frangible link was intentionally broke to terminate the tow. Tethered test points explored climb rate, descent, turns, the stable tow envelope in both clean and dirty configurations, and the effects of varying the EXD-01 drag.

5. RESULTS OF TESTS *(Continue on reverse if needed)*

The C-141 made a normal engine start and began taxi to the RWY 04 hammerhead at 1433 Z. There was a slight delay for the EXD-01 as the rope angle instrumentation was being examined. At 1456 Z the C-141 taxied onto the runway. At 1505 Z the rope was aboard the C-141 and 32,000 lbs of fuel remained. Slack removal procedures began at 1513 Z. Tension was set at 1520 Z. The flight was ready for takeoff at 1521 Z and chase immediately called 40 seconds. Brake release was at 1522 Z. For the desired thrust factor (TF) of 18.0, computed and set EPR was 1.92. Towed takeoff was normal.

Climbout was at 190 KIAS and power control was similar to previous flights. In the climb, the climb rate effects test points (trim shots) were taken at 1500 fpm and 1000 fpm rates of climb. The loadmaster reported the rope to appear very stable and that it was resting on the ramp edge in the climb. The flight leveled at 10,000' MSL at 1531 Z.

The flight proceeded with the stability boundary investigation (gear down, speedbrakes out) card at 1532 Z. The first half was performed at the lower tow position. Following this, at 1536 Z, the trim in turn card was completed in a right turn at bank angles of 15, 30, and 45 degrees. The EXD-01 reported marginal spiral stability in the turn. Following the turn, the stability boundary investigation card was returned to at 1540 Z for the upper tow position. Data was also taken at the point at which the tow rope appeared to be straight. The EXD-01 estimated this point to be at a declination angle of 6 deg. At 1544 Z the flight entered a turn to the left and took turn card data in the turn.

------ CONTINUES NEXT PAGE ------

6. RECOMMENDATIONS

While the rope configuration of today's flight was noticeably less damped without the nylon webbing in the tow train, handling qualities of the system remained quite satisfactory. Testing should proceed as planned.

Following today's flight, discussion occurred regarding upcoming flights, to include the flight to altitude (18,000 or 25,000 ft MSL). There appear to be some AFFTC safety concerns regarding the high altitude flight that will necessitate an additional amendment to the AFFTC safety package to address. This appears to primarily be a paperwork rather than an operational issue.

COMPLETED BY	SIGNATURE	DATE
Morgan LaVake, Test Conductor		23 JAN 98

AFSC Form 5314 NOV 86 REPLACES AFFTC FORM 365 MAR 84 WHICH WILL BE USED

Document 16. Daily/Initial Flight Test Report, C-141A, 61-2775, 23 Jan. 98, Morgan LaVake, Test Conductor

Section 5 continued.......

Out of the turn, the stability boundary card was again revisited at the upper position at 1548 Z. With the EXD-01 near the upper limit of the tow envelope, there was a very noticeable bungee mode. It could be observed visually in the rope and could be felt and seen in the C-141 aircraft response. The bungee oscillation caused a slight oscillation in pitch of the C-141 and also resulted in a noticeable longitudinal surging in the C-141. The pilot estimated the longitudinal accelerations around ± 0.1g with a 5 second period. A slight pitch oscillation accompanied the accelerations but was easily controllable by the pilot. Airspeed excursions were less than 2 knots. At the upper limit, the loadmaster reported the rope to be approximately 1.5 ft above the ramp edge.

This was followed by the drag effects on stability card at 1552 Z.

At 1555 Z the stability boundary effects card was began in the EXD-01 gear up, speedbrakes in configuration. C-141 fuel was 21,000 lbs. Once again, at upper tow position, there was very noticeable bungee mode that caused slight pitch oscillation in the C-141.

At 1612 Z the HQR Criteria - on tow card was performed. AS the EXD-01 changed configuration, bringing up the gear and retracting the speedbrake, the C-141 felt a slight forward surge.

At 1617 Z the upper tow limit / wake investigation card was performed. C-141 fuel load was 17,000 lbs. The loadmaster observed the rope to rise to an estimated 2.5 to 3 feet above the ramp edge.

At 1621 Z the flight descended to 5000' MSL and performed the stability in descent card in the descent.

The flight slowed to 170 KIAS for the tow release. The build up to the frangible link separation resulted in reported peak loads of 8,000 lbs, 13,000 lbs, and 21,500 lbs on successive maneuvers. This occurred at 1629 Z. At 1630:28 Z the frangible link was intentionally broken an the flight separated. The break maneuver was felt quite strongly in the C-141. The longitudinal deceleration was moderate and was described by the pilot as similar to moderate braking during a full stop landing. Due to short duration of the deceleration, the pilots did not observe a significant change in the airspeed. Upon frangible link separation, the rope was observed to significantly recoil, but no part of the rope recoiled into the C-141. Rope whipping of the free rope was significant, but similar to previous tests.

At 1636 Z the C-141 dropped the tow rope into the PB8 drop zone. The C-141 was at a minimum fuel state and as such the untethered test card was not performed.

The C-141 landed on Rwy 22 at 1644 Z.

The test pallet tape was provided to NASA for processing. The NASA Ashtec GPS was removed from the aircraft and returned to NASA for downloading. Postflight processing of the test pallet tape found it to be a satisfactory recording.

Aerodynamics
Eclipse
EXD-01 Flight #7 (third towed flight)
January 23, 1998

EXD-01 (NASA QF-106A 59-0130):   Stucky
C-141A (USAF C-141A 61-7775):   Farmer, Latimer, LeVake, Stahl, Brink, Drucker

Objectives for this test were to begin the entire test sequence over again, using the present priorities, with an all high tech liquid crystal polymer rope and no nylon damper in the tow train system. Priorities were: trim, dynamics, handling qualities, and rope characterization.

C-141: no change.

EXD-01: no change, with additional points added for clean aircraft characterization.

Tow train: no nylon in the middle, nominal 1000 feet of tow rope.

Chase support was provided by 846 (Jim/Lori) and 852 (Rog/Jim). No turbulence was noted, the conditions were cool (32F) and calm.

Lateral cable angle appeared wonky, and some time was lost in trying to characterize the problem (reduced dynamic range in lateral cable angle). The EXD-01 takeoff was at 07 22 45. Trim shots were performed at 1000 and 1500 fpm climb.

Stability boundary investigation showed much better comparison with the sim than in previous cases, though the flight is still more forgiving than the sim. Longitudinal instabilities (rope bungee mode) were found at high tow positions. Low tow poositions were characterized by lateral instabilities, in both bungee and spiral mode diovergence. In all cases, time to double amplitudes were on the order of 7-10 seconds, and the pilot was able to damp the motions out at will, only during stick fixed flight were the instbilities evident. Gimbals were called several times in the low tow positions. the card was repeated with the airplane clean, the instabilities were somewhat worse at high tow (longitudinal), and somewhat more benign at low tow (lateral). During this data MOF hits were noted.

HQR was performed again, ratings of 2 were given twice each in both dirty and clean configurations. Trim in turns was performed again, and 40 degrees angle of bank was noted as neutral spiral stability. Upper tow limit was performed again, com was garbled during this time. Stability in descent was done at 500 and 1000 fpm.

Document 17. Aerodynamics, Eclipse EXD-01 Flight 7, 23 January 1998, Al Bowers, Chief Engineer

The weak [frangible] link was intentionally broken in a build up fashion. Load peaks of 8k lbs, 13k lbs, 21.5k lbs were noted before breaking the weak link at 24,300 lbs at 08 30 31. The EXD-01 performed a landing at 08 47 08.

Take off time: 07 22 45
Release time: 08 30 31
Landing time: 08 47 08
Flight time: 01 24 23
Tow time: 01 07 46
Total Tow Time: 02 08 51

Al Bowers
Chief Engineer

Flight Controls
Eclipse
EXD-01 Flight #7
January 23, 1998

Brake release and takeoff.
Initial longitudinal acceleration was smooth, and slightly higher than during the takeoff of the previous flight (0.35 g s). Otherwise, the two takeoffs very similar.

Climb to 10,000 feet.
During the stabilized 1,500 ft/min climb a low-amplitude (less than –1/2 deg), pitch oscillation (with a period of appr. 3. 5 sec) was observed. Values of the tow rope tension were similar to those during Flight 6.

Stability boundary tests.
While probing the high and low boundaries, unstable bungee oscillations were encountered in response to both longitudinal and lateral-directional doublets at both extreme tow positions. However, the time to double-amplitude was relatively long, 7-8 seconds, so that the pilot could easily return to the stable tow positions. The longitudinal oscillations, set off by pitch doublets, coupled into roll oscillations, but the lateral-directional doublets resulted only in roll oscillations. As in the previous flight, the low tow position was limited by the 20-degree gimbal limit, while the upper position was limited by encountering the C-141A wake. Tow rope shapes during these tests were similar to those seen during the previous flight.

These tests were repeated in the clean configuration. It appeared that the stable tow region was narrower, and displaced higher than what was experienced with landing gear down and open speed brakes. However, in the stable region the clean airplane was almost completely free of the bungee oscillation with low values of tow rope tension (less than 5,000 lbs).

Drag cleanup tests.
No significant effects of the all-vectran tow rope were noted during these tests.

Turning flight.
Other than the negative 1-2 degree sideslip angles during the left turns at 30-degree bank angle, no significant differences from the results of Flight 6 were observed.

Descents on tow
The results of these tests were very similar to those of Flight 6.

Handling qualities tests.
No significant differences from the previous flight were noted during these maneuvers.

Document 18. Flight Controls, Eclipse Flight 7, 23 January 1998, Joe Gera

Tow rope release.
Instead of releasing the tow rope, the pilot of the EXD-01 airplane intentionally broke the frangible link at the planned flight condition. This was done by adding increasing amounts of power while still on tow to reduce tow rope tension, then chopping the throttle. No significant transients were noted during this maneuver.

Joe Gera

Structures Report
Eclipse
EXD-01 Flight #7
January 23, 1998

This flight was structurally severe (but OK) - intentionally reaching Design Limit tow rope load to break the frangible link demonstrating this release method and exploring the required maneuver loading technique. Parametric effects on tow stability were again studied. Load data quality was excellent including virtually zero offset. TM was generally good, but did have two time intervals (about 2.0 minutes and 1.5 minutes in length) during which annoying data spikes were present. This was thought to be a radiation pattern problem.

System configuration was significantly different for this flight. The center nylon damper segment was eliminated and the rope was 1,000 feet of continuous Vectran. This configuration produced less pronounced rope sail tendencies. Mid rope lift off load during tensioning was observed to be about 3,400 pounds (close to Bowers' prediction of 3,100 pounds). Previous configurations required more than 5,000 pounds of rope tension to lift the mid point off the runway. This matter is not a significant factor for operational procedures, etc. but was used as an additional tow rope load signal confidence (warm fuzzy) check for each flight. Primary load signal checks are the zero load offset and the RCAL. The absence of the 50 foot nylon section also produced a system that manifested lightly damped dynamic behavior that the control room heard called: "CLASSIC!" and "TEXT BOOK!". The C-141A Load Master made the real time call: "excitation on rope" during the most pronounced oscillatory event in which the loads varied from about 1,500 to 14,000 pounds at the extremes. The dynamic behavior shown on this flight was not a problem but did indicate the potential for slack or high load production which had been a consideration in earlier project safety planning.

Takeoff loads were moderate reaching 12,400 pounds at the third cycle peak and 12,700 pounds at about main gear lift off. No slack problems occurred during the take-off process.

The EXD-01 tow release was accomplished through the planned frangible link overload. The procedure involved the EXD-01 pilot adding power to generate slack while laterally offset, then reducing power to idle to pull back to create a peak load. This process was used four times in an incremental build-up producing peak loads of 9,000, 13,500, 21,800, and finally 24,300 pounds which successfully broke the frangible link and released the EXD-01 from towed flight. The EXD-01 then was free to proceed under powered flight and to land with the instrumented (for rope angles) knuckle unit still in the hook mechanism for reuse. The frangible link failure load for this flight was well within the nominal range of the 24,000 pound Design Limit load for which the QF-106 modifications and tow train were designed. Post flight aircraft inspections showed no structural yielding of any kind.

Bill Lokos,
Structures Engineer

Document 19. Structures Report, EXD-01 Flight 7, 23 January 1998, Bill Lokos, Structures Engineer

Eclipse
Weather Summary
EXD-01 Flight #7
January 23, 1998

High pressure at the surface and aloft was the dominant feature leading to the 3rd Eclipse
tow flight on January 23, 1998. The 24 hour forecast, issued the day before, called for
stable atmospheric conditions and cold overnight temperatures. The temperature forecast
for takeoff time of 0715 PST was 32 deg F. The surface winds were forecast to be
variable with speeds no greater than 6 knots. Light turbulence was forecast from surface to
8,000 feet. Sky conditions were forecast to be broken cloud coverage at 25,000 feet. No
precipitation or severe weather were forecast in the area.

Flight day weather observations were well within operational limits. The temperature at
takeoff time, which occurred at 07:23 PST, were observed to be in the low 30's. Winds
were westerly at with speeds ranging from 3 to 4 knots. The wind and temperature data
were observed from wind towers located along the main runway. Tower 044 is located
4000 feet down the threshold of runway 04 and Tower 224 is 4000 feet down runway 22.
Both towers measure data 30 feet above ground level. Sky conditions were observed to be
broken cirrus at 25,000 feet; right on forecast. No atmospheric turbulence or severe
weather were observed during the flight.

Casey Donohue

Document 20. Weather Summary, Eclipse, EXD-01 Flight 7, 23 January 1998, Casey Donohue

Eclipse-01	Flt Date	Flight No. F7

EXD-01 to C-141 Hook-up

KCAS	Altitude

1. **ECLIPSE** slow taxis into position - OFFSET upwind side ── *65734*
   (truck drives to centerline, 30 ft in front of hash marks)
2. Chock **ECLIPSE** wheels – *6 70019*

3. C-141 holds in position (middle of 1000 ft hash marks) with
   Before Taxi Check complete –
4. Handler remains at EXD-01 nose for rope management

5. Truck pays out tow rope to the C-141& stops at midpoint
   (FOD check while in transit) – *7:02*
5. Simultaneously C-141 inititates ERO checklist *7:02*

6. ARRIS 09 Transmits-"Cleared to approach" – *7:02*
7. Truck pays out remaining rope to the C-141
8. Remaining runway is FOD checked by NASA ground crew
   (simultaneously) *complete 7:07:16*
9. NASA ground crew connects the tow rope to the 3-pin
   connector *7:06:04*

(NEXT PT: _____)

A

Arris 09 Taxi - 65643

Eclipse Taxi - 65734

65734

702 - Clear to approach 09 (85)

Document 21. Example of flight test cards, Eclipse-01, Flight No. F7

Eclipse-01	Flt Date	Flight No. F7

Slack Removal

KCAS   **Altitude**

1. Ground crew clear runway at RWY 04 last chance
   (EXD-01 crew chief remains nearby)
2. ARRIS 09 performs ERO and Before Take-Off checks

3. Remove chocks from **ECLIPSE** — 7:13:34
4. ARRIS 09 transmits - "Ready to take up slack" – 7:13:34
5. **ECLIPSE** transmits - "Clear to take-up slack" – 7:13:34

6. ARRIS 09 begins slow roll to take-up slack (on mission
   frequency the load master will call "slack removed") – 7:13:36
7. ARRIS 09 continues to creep forward - 7:14:41
8. **ECLIPSE** monitors rope load and when reaching 2000
   lbs. (+/- 1000) transmits - "Hold Your Position" – ~~7:14:11~~ 7:15:28   7:15:15 1000
                                                                7:15:28 2000
9. ARRIS 09 holds brakes and transmits - "ARRIS 09
   Holding Position" — 7:15:40

10. **ECLIPSE** transmits - "**ECLIPSE** Holding Brakes" - 7:15:40
11. Crew chief clears area to last chance
12. ARRIS 09 verifies:
    -Guillotine safety straps are cut
    -Camera - ON
    -Intercom check complete
    -Line-up check complete (IFF - Standby)
    -TPS Pallet - "Data On"

(NEXT PT: _____)

B

Eclipse
EXD-01 Flight 8
Fourth Tethered/Release Flight
January 28, 1998

Project Manager's Comments

And I thought the last flight was perfect...this one was a true match, except we improved the timeline!

Today at the debrief I was watching the AFFTC team and was struck by their demeanor. They seemed to flow as a single entity. They had a calm look of capability. And that is the way I pereive them during the flight operations that we have conducted with them. Their roles and responsibilities are clear. They have a full understanding of the task, the plan, and the intended outcome. They are highly organized and precise in performing their tasks, but are flexible so that they can adapt to the changing situation during the actual mission. It's a good learning experience for me to work with such a high caliber group of people. Their participation is a major factor in why we are doing so well.

I think because things are going so well on our missions, that this is the time when I am supposed to dust off my "don't get too overconfident" lecture. I don't think I will, though, because it doesn't appear to me that this is happening. Instead, let's hope for continued favorable weather and keep getting good data! Good job, everybody!

Carol A. Reukauf

Document 22. Project Manager's Comments, Eclipse, EXD-01 Flight 8, Fourth Tethered/ Release Flight, 28 January 1998, Carol A. Reukauf

Eclipse
Pilot's Flight Test Report
EXD-01 Flight 8 - Fourth Tethered Flight, January 28, 1998

Overview
This flight built upon our previous flights with added emphasis on documenting the tow rope position. This was accomplished by observation of rope position by the loadmaster with simultaneous rearward and side view photographs (from the C-141A and F-18 chase aircraft).

Ground Operations
A change was made in the slack removal and tensioning procedures. The new procedures (which were pencil changes during the crew brief) meant that when slack was removed, no tension was set. Chocks were then installed on the EXD. Tension was set in one continuous step to 6000 lbs. The new procedures worked very well, and waiting for takeoff was no longer a pilot's isometric leg exercise. The timeline was adhered to very well and we were ready for takeoff at 0713.

Tethered Takeoff
The winds were 250° at 7 kts. The initial tension build-up after the brake release call was not as well controlled as on previous flights and peaked at approximately 18,000 lbs. Perhaps I wasn't watching the tension gauge as closely as before, or perhaps the C-141A crew added power / released brakes more abruptly (the pilot and copilot swapped roles on this flight). I damped the oscillations quickly and the rest of the takeoff was nominal. In retrospect, because the rope center section stays off the runway at >3000 lbs. of tension, there is no reason to try to maintain loads above 10,000 lbs. during the initial roll. For future takeoffs I **recommend using 6,000 lbs. as the initial targeted tension value immediately after EXD brake release.** Immediately after clearing the ground I closed the speedbrakes, waited several seconds for any potential bungee mode to damp and then raised the landing gear. It was an easy transition.

Flight Cards
Climb Rate Effect on Stability - Clean. The handling qualities seemed similar to the previous dirty tests.

Stability Boundary Investigation - Dirty. Changes from previous versions of this card included more discrete positions where doublets, as well as the simultaneous rope photography was done. The new procedures worked well.

Aircraft responses were stable at the medium position with an unstable dutch roll mode increasing at the lower limits. An unstable bungee mode developed in the upper position (just below the wake).

Stability Boundary Investigation - Clean. The EXD had stable responses in the medium and low positions. In the medium high position a lateral oscillation occurred. In the high position an unstable bungee coupled into the dutch roll.

Document 23. Eclipse Pilot's Flight Test Report, EXD-01 Flight 8—Fourth Tethered Flight, 28 January 1998, Mark P. Stucky, Eclipse Project Pilot

Eclipse
Pilot's Flight Test Report
EXD-01 Flight 8 - Fourth Tethered Flight, January 28, 1998

HQR Criteria - On Tow. All tasks were easy to relatively easy to perform. Tasks were easiest in the dirty configuration and at the nominal position. The task was slightly harder in the medium low tow position where responses felt less damped.

Similar responses were observed in the clean configuration with less restoring force causing a slightly higher workload. The HQR for the clean nominal tow position task was equivalent to the dirty and low task.

Trim in Turns - Clean Configuration. Turns were done both in the nominal tow and in the medium high (straight rope) positions. Hands free stability was achieved in turns of up to 45° bank. Approximately one click of nose up trim was required per 10° of bank angle.

Upper Tow Limit / Wake Investigation - Clean. I was able to climb to approximately 40 ft. above the C-141A where I reached the loadmaster's limit for rope to tail clearance. I flew through the vortex on the descent without any problems.

Lateral Offset - Clean. Doublets were performed from a fixed cross-controlled "neutral" point.

Stability in Descent - Clean. At the 1000 fpm descent rate the tension forces were very light (1,000 to 2,000 lb.). At these light values a rope "leaf spring" motion occurs which doesn't affect aircraft handling qualities or rope tension. **Recommend the maximum descent rate in the clean configuration be limited to 1000 fpm.**

Frangible Link Loads Excitation and Separation. There was no problem in the medium and medium low positions where the rope would smoothly bow away from the nose of the EXD. But in the medium high (straight rope) position, when the rope unloaded after the initial tension spike it suddenly transitioned from the smooth arc to an unnerving and chaotic spaghetti-like appearance. I immediately banked and descended to reestablish rope stability. I commented that I would not repeat that point again. **Recommend maintaining some tension anytime the tow rope is straight.**

Actual separation was uneventful.

RTB and Landing. Fuel permitted a climb to above 20,000 ft. and a gliding approach to a full stop landing.

Eclipse
Pilot's Flight Test Report
EXD-01 Flight 8 - Fourth Tethered Flight, January 28, 1998

Lessons Learned and Recommendations

1. Recommend using 6,000 lbs. as the initial targeted tension value immediately after EXD brake release.
2. Recommend the maximum descent rate in the clean configuration be limited to 1000 fpm.
3. Recommend maintaining some tension anytime the tow rope is straight.

Mark P. Stucky
Eclipse Project Pilot

DAILY/INITIAL FLIGHT TEST REPORT		1. AIRCRAFT TYPE	2. SERIAL NUMBER
		C-141A	61-2775

3. CONDITIONS RELATIVE TO TEST

A. PROJECT / MISSION NO	B. FLIGHT NO / DATA POINTS	C. DATE
Eclipse	F-8, 4th Towed Flight	28 JAN 98
D. LEFT SEAT *(Front Cockpit)*	E. FUEL LOAD	F. JON
Maj Kelly Latimer	35,000	C9703900
G. RIGHT SEAT *(Rear Cockpit)*	H. START UP GR WT / CG	I. WEATHER
Capt Stu Farmer	203,000 lbs / 28.8%	Clear
J. TO TIME / SORTIE TIME	K. CONFIGURATION / LOADING, SOFTWARE	L. SURFACE CONDITIONS
1514 Z / 1.4 hrs	Petal Doors Removed / Tow Config	28°F/ Winds 330 @ 5kts
M. CHASE ACFT / SERIAL NO	N. CHASE CREW	O. CHASE TO TIME / SORTIE TIME
F-18 NASA 846 & NASA 852		

4. PURPOSE OF FLIGHT / TEST POINTS

This flight was the fourth towed flight of the Eclipse program. The all Vectran rope was again used and the tow was terminated with an intentional break of the frangible link. Test maneuvers were very similar to the previous flight with the exception of the EXD-01 primarily performing cards in the gear up, speedbrakes in configuration, whereas the previous flight had predominantly been in the dirty configuration. Also performed on this flight were a series of simultaneously taken photographs from the C-141 on board photo and from the chase photo.

5. RESULTS OF TESTS *(Continue on reverse if needed)*

The C-141 made a normal engine start and began taxi to the RWY 04 hammerhead at 1440 Z. The EXD-01 was ready for taxi when the C-141 arrived, and the C-141 took the runway at 1447 Z. At 1456 Z the rope was aboard the C-141. Slack removal began at 1505 Z. The revised slack removal procedure was used in which the 2000 lb pre-load was not performed. This procedure worked well and is an improvement over the previous technique. Tension was set at 1512 Z. The flight was ready for takeoff at 1513 Z and chase was almost immediately in position. Brake release was at 1514 Z. For the desired thrust factor (TF) of 18.0, computed and set EPR was 1.92. C-141 fuel at takeoff was 32,500 lbs. Towed takeoff was normal.

Climbout was at 190 KIAS as in previous flights. In the climb, test cards were performed at 1500 fpm and 1000 fpm rates of climb. Simultaneous photos were taken at each climb rate and pitch and lat-dir doublets were performed by the EXD-01. The flight leveled at 10,000' MSL at 1521 Z.

The flight proceeded with the stability boundary investigation (gear down, speedbrakes out) card at 1521 Z. In this card, the EXD-01 performed longitudinal and lateral-directional doublets at multiple elevation positions relative to the C-141. At each position, simultaneous photos were taken prior to the doublet sets (no photo was taken at the lowest tow position as the EXD-01 was out of the photographer's field of view). During this card, a turn was entered, the EXD-01 reconfigured, and turn card data points taken. Following the turn, the stability boundary investigation was returned to. At the upper tow position, the bungee mode became quite evident upon the C-141, with significant pitch oscillations developing following the EXD-01's longitudinal doublet (1532 Z). At this point, the EXD-01 aircraft response appeared divergent and the EXD-01 recovered to a lower tow position.

------ CONTINUES NEXT PAGE ------

6. RECOMMENDATIONS

Proceed with planned testing.

Update test cards to reflect revised slack removal procedure (deletion of 2000 lb pre-load prior to setting tension).

Possibly increase drop airspeed from 140 KIAS to 150 KIAS to increase drag on the free rope and promote clean separation at rope drop.

COMPLETED BY	SIGNATURE	DATE
Morgan LaVake, Test Conductor		28 JAN 98

AFSC Form 5314 NOV 86 REPLACES AFFTC FORM 365 MAR 84 WHICH WILL BE USED

Document 24. Daily/Initial Flight Test Report, C-141A, 61-2775, 28 Jan. 98, Morgan LaVake, Test Conductor

Aerodynamics
Eclipse
EXD-01 Flight 8 (4th towed flight)
January 28, 1998

Flight Crew:
EXD-01 (NASA QF-106A 59-0130): Mark "Forger" Stucky
C-141A (USAF C-141A 61-7775):   Stu Farmer, Kelly Latimer, Morgan LeVake,
                                 John Stahl, Dana Brink, Ken Drucker

Eclipse Flight 8 (towed flight 4) took off at 07:14:49 and released at
08:23:00 with a landing at 08:36:38. Total flight time was 1 hour 21
minutes 49 seconds and total time on tow was 1 hour 8 minutes and 11
seconds. A tension spike occured right at brake release of 19,000 lbs
or so. Tow was terminated with a weak link break. The instrumented
knuckle appeared to be operable in the post flight check. Most of the
flight was flown in the clean configuration; the EXD-01 was noted to
have a decidedly different character clean. Reduced stability was
noted in some parts of the envelope.

Handling qualities (HQs) for the pilot degrade slightly due to
workload as the airplane cleans up. The HQR task used was lateral
offset. The EXD-01 lines up behind the C-141A number 4 engine, then the
pilot has to capture the centerline as quickly as reasonable. These
were performed at medium high tow position (the rope is straight is
the definition of medium high), a medium position ("nominal" tow
position), and the medium low position (where incipient instabilities
develop, or neutral stability). Desirable criteria are: No tendency
to aircraft-pilot-couple (APC), overshoot of less than 20 feet
(halfway between the number2/3 engine and the centerline of the
C-141A), and less than 2 lateral overshoots. Acceptable criteria are:
Any APC tendency is damped in less than 2 cycles, overshoots less than
40 feet, and less than 3 lateral overshoots.

This set of maneuvers was done at 10k ft msl and 190 KEAS. Dirty
medium-high HQR was rated a 2, dirty medium HQR was rated a 2, dirty
medium-low was rated a 2. The pilot required minimal compensation to
achieve the desirable task. Comments from these points (in order)
were: "no tendencies to APC, no overshoots"; "No overshoots, no
tendencies to APC, but it required more cross control to maintain
position"; and finally "Desirable performance, overshoot of about 15
feet; it is less damped here." The points were repeated with the
EXD-01 clean and the ratings were 2 for medium-high, 3 for medium, and
3 for medium-low. The comments recieved were: "overshoot of about 15
feet, equivalent the dirty med-lo point"; "the same, but more active",
and finally "no overshoots, no tendency to APC, but increased

Document 25. Aerodynamics, Eclipse EXD-01 Flight 8, 28 January 1998, Al Bowers, Chief
Engineer

Section 5 continued.......

At 1534 Z the stability boundary effects card was repeated in the clean configuration. Again, simultaneous photos were taken at multiple tow elevations. Additional turning card test data was taken interspersed into this card.

This was followed by the HQR Criteria - on tow card at 1547 Z.

The upper tow limit / wake investigation card was then performed. During this maneuver, the EXD-01 rose to an elevation well above that seen on previous tethered flights. The loadmaster made an advisory call to the EXD-01 not to go higher, as the rope was estimated to be 2-3 feet from contacting the upper empennage of the C-141.

The lateral offset - clean test card was next performed, at 1601 Z. C-141 fuel was 20,000 lbs.

The TPS Instrumentation Pallet data tape was changed at 1604 Z prior to the next card.

At 1607 Z the flight descended to 5000' MSL and performed the stability in descent card in the descent.

The flight remained at 190 KIAS for the tow release by frangible link separation. Load maneuvers were 1st performed by the EXD-01, at nominal, med-high, and med-low tow positions. During these load maneuvers, the C-141 airspeed was observed to increase 2-3 knots during the momentary slack rope period. Upon application of the EXD-01 load, the airspeed was observed to remain at the new higher airspeed. The frangible link was broken by the intended release maneuver at 1622:57 Z.

At 1627 Z the C-141 attempted to drop the tow rope into the PB8 drop zone. The primary and secondary guillotines were activated, but the rope remained attached to the C-141. As the loadmaster was moving aft to investigate the malfunction, the rope was observed to depart the aircraft as the C-141 entered a turn. Approximate drop coordinates as indicated by the INS were N34°52.78' x W117°40.68'. Postflight inspection revealed that the nylon webbing was not completely severed by the guillotine at the edge of the webbing.

The untethered test card was not performed.

The C-141 landed on Rwy 22 at 1633 Z.

The test pallet tape was provided to NASA for processing. The NASA Ashtec GPS was removed from the aircraft and returned to NASA for downloading.

workload." This qualitative assessment of the aircraft flying qualities is consistent with the stability data gathered from the doublets.

Flight 8 (fourth towed flight)
Date: 01/28/98
Take off Time: 07 14 49
Release Time: 08 23 00
Landing Time: 08 36 38
Flight Time: 01 21 49
Tow Time: 01 08 11
Total Tow Time: 03 17 02

Al Bowers
Chief Engineer

Eclipse
Flight Controls
EXD-01 Flight 8
January 28, 1998

Brake release and takeoff.
Brake release was followed by an abrupt longitudinal acceleration of 0.45 g's resulting in a peak tow rope tension of just under 19,000 lbs. Speed brakes and landing gear were retracted immediately after liftoff. No transients were noted during the transition.

Climb to 10,000 feet.
The climb was made at 1,500 and 1,000 ft/min climb rates with doublets performed about each axis. No significant effect of the two climb rates on the responses was noted.

Stability boundary tests.
Gear down, speed brakes out. At the lower limit of stability a lightly damped, but still stable pitch oscillation was observed. However, the lateral-directional doublet resulted in an unstable dutch roll oscillation with a time to-double-amplitude of approximately 10 seconds. At the larger amplitudes kinematic coupling induced an alpha oscillation at twice the frequency of the dutch roll. The higher limit of stability was defined by an unstable longitudinal bungee oscillation with a time to-double-amplitude of approximately 4 seconds. A lateral-directional doublet coupled into the longitudinal bungee mode. Despite the high degree of instability at both the higher and lower tow positions the pilot could readily move a stable tow position. He stated that the unstable modes and the techniques to suppress them reminded him of the Eclipse simulator. Gear up, speed brakes closed. No instabilities were encountered down to the lowest tow positions possible without bumping into the low gimbal limit. While probing the higher stability boundary in the medium-high tow position, an unstable lateral rope oscillatory mode with a relatively long period (T ~ 13 sec) was observed. The high tow position resulted in the longitudinal bungee mode that was less unstable than that in the landing configuration.

The upper tow limit was probed in the clean configuration. The highest position reported by the pilot was about 40 feet above the C-141 wing wake. The EXD-01 remained completely controllable in the high tow position in the center, and in a laterally offset position during which the wing tip vortex from the tow plane was briefly encountered.

Turning flight.
At various times throughout the flight turns at bank angles of 15, 30, and 45 degrees were set up by the C-141 tow plane. As in previous flights the EXD-01 had no difficulties while following the tow plane in turns. In fact, even in the highest bank angle of 45 degrees it was possible to trim the EXD-01 for hands-off flight.

Document 26. Flight Controls, Eclipse Flight 8, 28 January 1998, Joe Gera

Descents on tow.

Rates of descent of 500 and 1,000 ft/minute were set up by the tow plane with the EXD-01 in the clean configuration. There was no tendency for tow rope slack during these descents, although the tow rope tension was reduced to very low values, approximately 2800 lbs, at times. During these low values of the tension, the tow rope was either completely quiescent or oscillating by itself without affecting either the tow plane or the EXD-01.

Handling qualities tests.

From lateral offsets, both in the clean and dirty configurations, the pilot aggressively flew back to the center to capture the nominal tow position. The pilot commented that the task was somewhat easier in the dirty, i.e., gear down and speed brakes out, configuration. The Cooper-Harper ratings were 2 and 3 for the dirty and the clean configurations, respectively.

Tow rope release.

The tow rope release for this flight was preceded by maneuvers designed to set of a longitudinal oscillation of the tow rope. This was done by adding power by the EXD-01 until the tow rope tension was reduced to very low values, and then chopping throttle. During one of these maneuvers the tow rope developed a substantial slack and bundled up in front of the EXD-01 forcing the pilot to perform a quick evasive maneuver.

Instead of using the tow release system, tow rope separation was effected by intentionally braking the frangible link. The pilot of the EXD-01 added power while in a lateral offset until a slack developed. A rapid deceleration resulting from a throttle chop and speed brake deployment created enough tension to break the frangible link so that the release hardware remained attached to the EXD-01, and thus could be used again for a subsequent flight.

Joe Gera

Flight Mechanics
Eclipse
EXD-01 Flight #8
January 28, 1998

All flight 8 data has been successfully processed, and removal of time skews and parameter biases is ongoing. Some initial results are worth reporting, however. The "stability boundary" test points yielded preliminary flight-determined trim curves for the towed configuration in both the "clean" and "dirty" configurations. The "high unstable" test point in the "dirty" configuration yielded a flight-determined long-elongation curve that was in reasonable agreement with the ground test data collected at the TMT test lab. The pair of differentially-corrected GPS receivers in both aircraft proved to be fully capable of extracting this unique piece of flight test data.

The process for documenting the rope sail via photography was fairly successful on flight 8. Approximately half of the phots taken from the chase aircraft appear to be useable, and almost all of the shots taken from the C-141A appear to be useable. The photos have yet to be measured, however. Flight 8 also provided critical experience for the photographers.

Jim Murray

Document 27. Flight Mechanics, Eclipse, EXD-01 Flight 8, 28 January 1998, Jim Murray

Structures Report
Eclispe Flight 8
Fourth Tethered Flight
January 28, 1998

This flight produced the highest tow rope loads to date.

System configuration was similar to last flight (one piece Vectran rope with no nylon damping segment) with the exception of several more painted tape marker segments.

Load signals zero load offsets were small. The take-off roll tow load profile was similar in shape to the previous case but produced a magnitude of about 19,000 pounds (more than 50% higher) in the first peak. This was very close to the threshold for the Red Light call at 20,000 pounds. Frangible Link yield begins around 21,000 pounds. This was not a problem but came close to aborting a good flight. As anticipated the all Vectran system tends to produce greater load extremes.

A repeat of a longitudinal doublet with gear down and speed brakes out at 10,000 MSL produced the highest maneuvering tow load reaching about 17,500 pounds. Some maneuvers produced rope waves large enough, although at very low loads, to exceed the 20 degrees Gimbals elevation limit.

Rope sail was observed and photographed at many different conditions by the photo chase. Video coverage, ground and chase, was again excellent.

During the beginning of the last tethered flight card a significant slack event was produced, which while well controlled by the pilot, presented some risk of entanglement with the pitot probe.

Having practiced the technique for intentional Frangible Link failure during the previous flight only one attempt was necessary this time producing an abrupt load peak reaching about 25,000 pounds neatly breaking the weak link and releasing the Tow Rope. Post release by the EXD-01 the Tow Rope flapped relatively mildly behind the C-141 until it was cut loose. The Guillotine operation apparently failed (both blades) to fully cut all of the fibers of the nylon straps resulting in several seconds of delay while the remaining nylon strands failed due to the light drag load on the mildly flailing rope.

Bill Lokos, Structures Engineer

Document 28. Structures Report, EXD-01 Flight 8, 28 January 1998, Bill Lokos, Structures Engineer

ECLIPSE
EXD-01 Flight 8
Weather Summary
January 28, 1998

Surface and upper-level high pressure with a weak offshore flow over Southern California was the situation leading to Flight 8 on Wednesday, January 28, 1998. Cold and stable weather was forecast to persist on flight day with variable winds with low or no turbulence. Sky coverage was forecast to be scattered to broken at 20,000 feet. Overall conditions were expected to be good for the flight.

Flight day observations were very close to forecast. The minimum temperature observed by the runway wind towers ranged from 28 deg F to 31 deg F. The wind speeds did not exceed 5 knots during the ground and takeoff operation. The winds observed at takeoff time, at 07:14 PST, were northeasterly at 3 knots. Since the winds were light and along runway 04, there were not significant crosswinds during takeoff. The wind and temperature data were observed from wind towers located along the main runway. Tower 044 is located 4000 feet down the threshold of runway 04 and Tower 224 is 4000 feet down runway 22. Both towers measure data 30 feet above ground level. No turbulence was reported during the flight. Cloud coverage was observed as broken at 25,000 feet. No severe weather or precipitation was observed in the area.

Casey Donohue

Document 29. Eclipse, EXD-01 Flight 8, Weather Summary, 28 January 1998, Casey Donohue

Eclipse
EXD-01 Flight 9
Fifth Tethered/Release Flight
February 5, 1998

Project Manager's Comments

Whew! What a flight! This mission was unique in that it included several added non-research objectives as well as the Eclipse objectives. On top of that, the effort to achieve the research objectives was in itself complex.

Today was Eclipse's Media Day. This was no small effort, but it was smoothly accomplished by our joint DFRC/KST media team made up of Fred Brown and Gray Creech on the NASA side and Emily Chase on the KST side. We had good media representation, everyone was pretty impressed by our project activities, and I consider the event to have been a major success!

The other unusual activity was that a helicopter film crew used this mission to obtain the footage that will become the documentation of our efforts for future national film presentation of aerospace accomplishments. Wolfe Air Aviation, a professional company that specializes in both commercial and entertainment productions, provided the helicopter, pilots, and camera personnel. Jim Ross and Lori Losey provided the initiative and contact. Charles McKee and Brent Wood handled the contracting. Forger provided the planning and organizing interface to the project. Marta Bohn-Meyer and Lee Duke obtained the USAF approval. And Roy Surovec, Stu Farmer, and Bob Wilson obtained the actual signatures. Nearly a cast of thousands! It was a very impressive effort, and it looks like the payoff will give us great pride that our efforts on Eclipse were an aviation milestone.

The most significant of today's effort was the pulling together all of the elements that will enable Jim Murray to better understand the "rope sail" phenomenon. The tiny little load cell that was inserted into the tow rope at the C-141A end was the culmination of several major efforts. The unit was fabricated in the machine shop, the portable computer system was built up in the Thermal Lab by Mark Nunnelee (who also managed to save the day during the mission as the computer operator flying on the C-141A), and the entire system was end-end tested in the C-141A on the ground. Meanwhile, the tow rope was marked and qualification tested. And at the same time inflight procedures for photographing the rope were being developed. The most significant aspect of this very complicated effort was that it was just conceived during the past three weeks since we learned that the behavior of the tow rope in flight was different than predicted. This is a superb example of what Dryden does well. And another fine example of the payoff of conducting flight research -- "...separating the real from the imagined..."

Document 30. Project Manager's Comments, Eclipse, EXD-01 Flight 9, Fifth Tethered/Release Flight, 5 February 1998, Carol A. Reukauf

These efforts were all brought together and integrated into our normal flight routine by our Operations Engineer and Mission Controller, Mark Collard, and the two flight crews, particularly Forger and Stu Farmer.

It was all exceedingly well planned and extremely impressive in its execution.

Congratulations to everyone involved!

Carol A. Reukauf

Eclipse
Pilot's Flight Test Report
EXD-01 Flight 9 - Fifth Tethered Flight, February 5, 1998

Overview
This flight had special emphasis on documenting the tow rope position as well as the load at both ends of the tow rope. A load cell was added to the C-141A three-pin connector using the standard end fitting used on the EXD end. The output was routed to a portable computer which was operated by a NASA - DFRC specialist. Additionally, we had a civilian helicopter for filming the ground operations and takeoff. This required special coordination for aircraft deconfliction, to mitigate potential hazards, and ensure all involved would know their roles and responsibilities. All went very well, although the helicopter rotors did blow the rope around a bit on his first pass and blew some sand on the runway during the takeoff. Airfield management was prepared for the runway FOD and immediately closed and swept it after the Eclipse flight tookoff. Lastly, we took off from Runway 22 because of the forcasted winds. This required the additional coordination to have the military security police close Lancaster Blvd. to minimize any potential hazards to ground traffic.

The flight cards did not flow as smoothly as on previous flights because of the photo documentation requirements and the broken overcast which caused delays while the flight repositioned so the rope would be visible in the photographs.

Ground Operations
Because the EXD had a single UHF radio and the helicopter was VHF only, we used two mission frequencies. The C-141A, NASA 1, and chase F-18s monitored both. Additionally, we had Chase #2 use the F-18's automatic radio relay capability, the end effect being that the EXD and helicopter could monitor and talk to each other as if they were on the same frequency.

We used very similar procedures as used on Flight 8 (altered slightly due to film requirements). A "20 second" call was added which was broadcast by NASA 1 twenty seconds after the brake release call. This call meant the helicopter had to transition from a head-on low altitude location 8000 ft in front of the C-141A's initial position to a position clear of the runway with a 200 ft. lateral offset.

Tethered Takeoff
Tension control and takeoff were uneventful.

Flight Cards
Climb Rate Effects of Stability - Clean. This card used a 2000 fpm climb rate which was visibly different (steeper climb angles) than the lower climb rates used on previous flights. I did not notice any degradation in handling qualities.

Stability Boundary Investigation - Clean. I had a natural tendency to level off with the same trim value used during the climb and settle into a lower than nominal tow position. Thus while the nominal climb position may have been

Document 31. Eclipse Pilot's Flight Test Report, EXD-01 Flight 9—Fifth Tethered Flight, 5 February 1998. Mark P. Stucky, Eclipse Project Pilot

Eclipse
Pilot's Flight Test Report
EXD-01 Flight 9 - Fifth Tethered Flight, February 5, 1998

"medium", after level off I was medium low and when a lower tow position was called for I hit the gimbal limits. Therefore, step 2c (low unstable) was not done since step 2b was already there.

Response to doublets was stable at the lower positions. At the medium high position the lat/dir doublet resulted in a dutch roll oscillation. Instability increased in the high position and the unstable dutch roll was now excited by the pitch doublet. It took a couple of attempts to reach the high tow (above the wake) position (vortex encounters would drop me back below). In this extreme high position the EXD was level with the top of the C-141A's T-tail. Unstable rope oscillations were prevalent, and I was forced to recover immediately after initiating the lat/dir doublet because of the proximity of the pitot tube and rope.

Trim in Turns - Clean Configuration. Changes on this flight included a turn series in the medium high tow position (with a straight rope) as well as a 15° to 15° banked hands off turn reversal in both tow positions. During the hands off turn reversals the EXD lagged the C-141A. In the nominal position a dutch roll oscillation occurred but was damping out. In the straight rope tow position the dutch roll oscillation was unstable.

C-141A Control Input Effects. The C-141A used a build-up approach and although the initial doublets were small, I could see the rope oscillation and feel a response. On the second (larger) pitch doublet the EXD climbed approximately 75 feet, starting a phugoid type response of relatively short period. The lat/dir inputs caused altitude changes also as well as a dutch roll response. When the slow pitch oscillation was performed tension values at the EXD end cycled from 1500 to 7000 lb. and the rope oscillated +/- 20 ft.

Stability Boundary Investigation - Dirty. Response was stable in the nominal position. As I descended the pitch response became very lightly damped in the medium low and coupled to a small dutch roll oscillation with an approximate 3.5 second period at the low position. As I climbed above the straight rope medium high position the bungee mode would be excited and resulted in quick termination and recovery. Attempting to get to high tow resulted in full aft stick and wallowing in the wake. Control authority was limited enough that doublets were not performed, but I held stick fixed and an unstable oscillation quickly developed (approx. 2.5 second period).

Stability in Descent - Clean. At the 1000 fpm rate of descent the EXD felt looser in all axis. Opening and closing the speed brakes caused a large rope oscillation which required actively using the speedbrakes (opening and closing them) to dampen the motion.

We reached the 5000 ft MSL level off altitude and crossed over the North base area on a heading for the PIRA.

Eclipse
Pilot's Flight Test Report
EXD-01 Flight 9 - Fifth Tethered Flight, February 5, 1998

Frangible Link Separation.
There were no buildup maneuvers and the separation was uneventful.

HQR Criteria - Off Tow. The capture tasks were relatively easy with HQR ratings of "3" due to the increase in workload to aggressively capture the centerline. I was rushing because of the fuel and did not know (understand) the requirement for stabilized flight on the centerline of the C-141A prior to offsetting and performing the HQR task. This wasted time and still did not achieve the quality of data the researchers desired.

EXD L/D Measurement. Done per the flight card.

RTB and landing was uneventful.

Mark P. Stucky
Eclipse Project Pilot

DAILY/INITIAL FLIGHT TEST REPORT		1. AIRCRAFT TYPE	2. SERIAL NUMBER
		C-141A	61-2775

3. CONDITIONS RELATIVE TO TEST

A. PROJECT / MISSION NO	B. FLIGHT NO / DATA POINTS	C. DATE
Eclipse	F-9, 5th Towed Flight	05 FEB 98

D. LEFT SEAT *(Front Cockpit)*	E. FUEL LOAD	F. JON
Maj Kelly Latimer	38,000	C9703900

G. RIGHT SEAT *(Rear Cockpit)*	H. START UP GR WT / CG	I. WEATHER
Capt Stu Farmer	206,000 lbs / 28.8%	Clear

J. TO TIME / SORTIE TIME	K. CONFIGURATION / LOADING, SOFTWARE	L. SURFACE CONDITIONS
1535 Z / 1.7 hrs	Petal Doors Removed / Tow Config	39°F/ Winds 210 @ 10kts

M. CHASE ACFT / SERIAL NO	N. CHASE CREW	O. CHASE TO TIME / SORTIE TIME
F-18 NASA 846 & NASA 852 + Helo		

4. PURPOSE OF FLIGHT / TEST POINTS

This flight was the fifth towed flight of the Eclipse program. The all Vectran rope was again used and the tow was terminated with an intentional break of the frangible link. As in the previous flight a series of simultaneously taken photographs from the C-141 on board photo and from the chase photo were taken. Slack removal and takeoff for this flight were filmed from a helicopter. This flight differed from the previous flight primarily in that a load cell was installed to measure and record rope tension at the C-141 end of the tow train. Additionally, this flight included a new test card, C-141 Control Input Effects, in which the C-141 performed doublets and a pitch oscillation with the EXD-01 on tow.

5. RESULTS OF TESTS *(Continue on reverse if needed)*

The C-141 made a normal engine start at 1436 Z and began taxi to RWY 22 at 1442 Z. Upon arrival at the last chance area, there was a slight delay while a fluid dripping from the C-141 was investigated. The fluid was found to be water, it had apparently pooled in the aircraft after heavy rains in the last several days.

The C-141 took the runway at 1455 Z. At 1507 Z the rope was aboard the C-141. Slack removal began at 1518 Z. The helicopter was photographically documenting the hookup and slack removal procedure. Slack removal generated an approximately 2000lb tension and the helicopter was satisfied with the video shot. The EXD-01 then released brakes and tension dropped. There was a hold for several minutes waiting for the scheduled 1530 Z takeoff time to approach. This was dictated by the runway 22 takeoff and pre-coordinated closure of Lancaster Boulevard to minimize the overflight hazard during takeoff. Additional checklist items were completed and tension was set at 1532 Z. The flight was ready for takeoff at 1533 Z and chase was immediately in position. Brake release was at 1534 Z. For the desired thrust factor (TF) of 18.0, computed and set EPR was 1.93. C-141 fuel at takeoff was 33,000 lbs. The photo helicopter was located over the runway at the 9000' point and filmed the takeoff. Towed takeoff was normal.

Sometime during the takeoff roll, the C-141 load cell instrumentation stopped functioning. Initial diagnosis was that the quick release connector had become undone during the takeoff. However, inspection showed it to remain connected. After approximately 10 minutes, the system was restored by the NASA instrumentation technician on board and data was recorded for the remainder of the flight.

Climbout was at 190 KIAS as in previous flights. In the climb, test cards were this time performed at 2000 fpm. The EXD-01 noted that this increased climb rate was distinctively a climb, unlike the lower 1000 and 1500 fpm climbs. Simultaneous photos were taken and pitch and lat-dir doublets were performed by the EXD-01 in the climb. The flight leveled at 10,000' MSL.

------ CONTINUES NEXT PAGE ------

6. RECOMMENDATIONS

Proceed with planned testing.

COMPLETED BY	SIGNATURE	DATE
Morgan LaVake, Test Conductor		05 FEB 98

AFSC Form 5314 NOV 86 REPLACES AFFTC FORM 365 MAR 84 WHICH WILL BE USED

Document 32. Daily/Initial Flight Test Report, C-141A, 61-2775, 05 Feb. 98, Morgan LaVake, Test Conductor

F9

Section 5 continued.......

There were several minutes of delay due to poor sun conditions for the chase photo. The C-141 control inputs card was next performed. The C-141 performed longitudinal and then lateral/directional doublets. This was followed by the C-141 performing a pitch oscillation with an approximately 5 second period. Maximum pitch oscillation of the C-141 was noted to be approximately ±2 degrees from the straight and level pitch attitude. Handling qualities of the C-141 were not perceived to differ appreciably from the not-on-tow handling. Both the doublets and the pitch oscillation developed rope oscillations that were experienced by the EXD-01.

Testing proceeded with the stability boundary investigation (gear up, speedbrakes in) card at 1553 Z. In this card, the EXD-01 performed longitudinal and lateral-directional doublets at multiple elevation positions relative to the C-141. At each position, simultaneous photos were taken prior to the doublet sets. In the highest tow position, the EXD-01 was out of the C-141 photographer's field of view. Interspersed into this card and later throughout the flight were turns during which data was taken. Assorted turns to both the left and right were accomplished at bank angles of 15, 30, and 45 degrees.

At 1610 Z the stability boundary effects card was repeated in the dirty configuration. Again, simultaneous photos were taken at multiple tow elevations. At high tow elevations, a bungee mode developed that caused slight pitch oscillations and a notable but slight longitudinal surging of the C-141 as seen in previous flights. When the EXD-01 positioned at the highest tow position, the C-141 pilot noted that it required a power increase to maintain airspeed.

The TPS Instrumentation Pallet data tape was changed at 1624 Z prior to the next card.

At 1627 Z the flight began a descent to 5000' MSL and performed the stability in descent card in the descent.

The flight remained at 190 KIAS for the tow release by frangible link separation. This time, no build up to the break event took place. The frangible link was broken by the intended release maneuver at 1638:42 Z.

The C-141 continued and released the rope into the PB8 drop zone. The primary and secondary guillotines were activated in close succession, and the rope departed the aircraft upon activation of the secondary release.

The C-141 climbed to 10,000' MSL and the EXD-01 rejoined for the HQR Criteria - Off Tow test card. This card was accomplished in the clean configuration only.

The C-141 returned to EDW for landing but was orbited by approach for approximately 5 minutes apparently due to a full pattern and busy tower controller. The C-141 landed on Rwy 22 at 1712 Z with approximately 14,500 lbs of fuel remaining.

The test pallet tapes were provided to NASA for processing. Of note is that the TPS Pallet ITAS computer, the ITAS video monitor, and the strip chart recorder were removed from the test instrumentation pallet prior to this flight in preparation for the unpressurized flight to 25,000' MSL. Instrumentation technicians did however confirm that the data recorder remains fully functional with these components removed. On the flight deck it was noted during the flight that the test instruments in the panel (alpha, beta, and N_2) were not functioning. The NASA Ashtec GPS was removed from the aircraft and returned to NASA for downloading.

Aerodynamics
Eclipse
EXD-01 Flight 9 (5th towed flight)
February 5, 1998

Flight Crew:
EXD-01 (NASA QF-106A 59-0130): Mark "Forger" Stucky
C-141A (USAF C-141A 61-7775):   Stu Farmer, Kelly Latimer, Morgan LeVake,
                                 John Stahl, Dana Brink, Ken Drucker

A load cell was added as instrumentation to the C-141A. Nearly identical
data sets are at both ends of the rope, with the exception of the rope angles at
the EXD-01 end.

The take off was nominal, and the cruise of 10k ft msl and 190 KEAS was
established.

The tow rope was marked every 100 feet with tape and paint to photograph
the rope sail; this documentation should assist in the assessment of static
trim effects of rope sail. This flight would be the only chance at getting good
rope sail data.

Also investigated was extreme high tow (about 70 feet above the C-141A) and
instability there. Doublets were out of the question to try and characterize
the dynamics; instead the pilot held the stick fixed for as long as possible.
The dynamic oscillations quickly built the tow tensions to about 18,000 lbs
and the EXD-01 was recovered.

Because the data set was so complete at both ends of the tow train,
maneuvers were also performed by the C-141A crew for measurement at
the EXD-01 end.

A weak link break was performed for tow separation. The C-141A dropped
the entire tow train assembly (including the load cell in the tow rope
assembly at the C-141A end) and both aircraft recovered on the runway.

Flight 9 (fifth towed flight)
Date: 02/05/98
Take off Time: 07 35 14
Release Time: 08 38 43
Landing Time: 09 01 16
Flight Time: 01 26 02
Tow Time: 01 03 29
Total Tow Time: 04 20 31

Al Bowers
Chief Engineer

Document 33. Aerodynamics, Eclipse EXD-01 Flight 9, 5 February 1998, Al Bowers, Chief Engineer

Eclipse
Flight Controls
EXD-01 Flight #9
February 5, 1998

Brake release and takeoff.

Brake release was smooth, peak longitudinal acceleration remained below 0. 3 g's with approximately 13,500 lbs of tow rope tension.

Climb to 10,000 feet.

The climb to the target altitude was made at 2,000 ft/min. Three axis doublets were performed during the climbing flight. While no significant climb rate effects on the responses were evident, the pilot commented after the flight that the climb rate was noticeable on his relative geometry with the tow plane.

Stability boundary tests.

Gear up, speed brakes in.   Because of gimbal limit, the low-unstable tow position could not be reached. In the high-unstable position the longitudinal bungee mode was neutrally damped. The EXD-01 airplane was flown through the wake to the high tow position. The response of the airplane to a pitch doublet consisted of a coupled longitudinal/lateral-directional oscillation that was neutrally damped. During the response to a lateral-directional doublet the tow rope was uncomfortably close to the pitot probe, so that the maneuver could not be fully investigated.

Gear down, speed brakes out.   In the lowest tow position a very lightly damped dutch roll motion was exhibited be the airplane. In the high-unstable position a longitudinal unstable bungee mode was observed in response to both the pitch, and the lateral-directional doublets. No doublets were performed in the tow position above the wake; the pilot commented that in that position he reached the reached the limit of controllability. He also commented on a drop of the airspeed of the C-141.

Turning flight.

In the turns during this flight a slight amount of negative sideslip (less than 2 degrees) was observed and removed by the pilot by applying of nose-left trim. As in the previous flight, the airplane could be trimmed for hands-off flight in the turns and turn reversals; however the pilot noted that during the reversals the EXD-01 lagged the tow plane, and that his workload was higher in the straight-rope tow position.

Descents on tow.

During the 500-ft/min descent in the lower tow position, the EXD-01 was noticeably looser. When the rate of descent was increased to 1,000 ft/min the tow rope tension reduced to slightly over 1,000 lbs, and then began a large-amplitude oscillation. The pilot modulated the speed brakes to damp out these, and commented that the 1,000 ft/min descent appeared to be the upper limit in the clean configuration.

Document 34. Eclipse Flight Controls, EXD-01 Flight 9, 5 February 1998, Joe Gera

Handling qualities tests.
The handling qualities task of the previous flight was repeated off tow, behind the C-141A. The control inputs and the resulting angular velocities were similar to the on-tow maneuvers of the previous flight. A pilot rating of 3 was given to the task.

Effects of C-141A control inputs.
While the EXD-01 was in the clean configuration, the C-141A performed small amplitude doublets about the pitch, and then about the lateral-directional axes. These resulted in slow, small amplitude pitch and dutch-roll type oscillations, respectively. When the C-141A pilot performed oscillatory pitch inputs with a 5-second period, the tow rope response was a large amplitude, leaf-spring type oscillation.

Tow rope release.
The frangible link was separated over the PIRA in a manner similar to that used in the previous flight.

Joe Gera

Flight Mechanics
Eclipse
EXD-01 Flight #9
February 5, 1998

Flight 9 had the unique addition of the load cell at the C-141A. Although some of the load cell data was lost due to connector problems, about 80 percent of the flight was recorded. The critical phases of the operation, including slack removal, tension setting, and the trim points, were all successfully documented. After some difficulty, the C-141A load cell data was synchronized with the EXD-01 PCM data. Initial inspection of the load data shows the difference in rope tension between the C-141A attach point and the EXD-01 attach point to be in fair agreement with current "rope sail" models.

Photo documentation of the rope from photo chase and the C-141A was made as part of the flight cards. Results from these efforts are not known at the time of this report.

Jim Murray

Document 35. Flight Mechanics, Eclipse, EXD-01 Flight 9, 5 February 1998, Jim Murray

Structures Report
Eclipse
EXD-01 Flight #9
February 5, 1998

System configuration was similar to last flight with the exception of the addition of a load cell between the three pin connector and the Vectran rope at the C-141A. This 30,000 pound range load cell was the one originally selected for use at the EXD-01 end but later replaced by the calibrated instrumentation added to the frangible link. The rope was 1000 continuous feet of Vectran with many painted tape marker segments and no center nylon strap damper segment.

Frangible link load signal zero load offsets were small. The takeoff roll tow load profile was similar in shape to previous cases and produced a moderate maximum magnitude of about 13,500 pounds, which occurred in the first cycle peak after EXD-01 brake release. No slack occurred during take off.

The C-141A load cell signal was recorded and displayed real-time by a modified lap top computer monitored by a technician seated at the video station forward of the tow tub. This miniature data acquisition system used a non-flight quality signal processing card which malfunctioned immediately after takeoff. After quick troubleshooting data was restored. This malfunction was reported to have recurred two subsequent times later in the flight and was quickly corrected each time. These events validated the decision to place a load measurement technician on the aircraft. It is estimated that about 95% of the tow rope load at the C-141A end was captured and recorded. These data were down-loaded, plotted, and transferred to other engineers the afternoon of the flight day. Post-flight inspection of the recovered tow rope assembly showed that the C-141A load cell assembly appeared undamaged by the rope drop procedure.

The peak maneuver load of about 16,000 pounds occurred during a longitudinal doublet in high (unstable) tow with a dirty configuration at 10,000 feet altitude. The second highest maneuver load (14,500 pounds) occurred during a lateral/directional doublet at the same condition.

An uneventful release was accomplished by an intentional frangible link failure which produced a single load peak to 25,000 pounds. This allowed returning with the knuckle assembly for reuse.

Bill Lokos

Document 36. Structures Report, EXD-01 Flight 9, 5 February 1998, Bill Lokos

Eclipse
EXD-01 Fight 9
February 5, 1998
Instrumentation Status Report

1) New calibrations for this flight:

TOWLDP - Tow Load Primary
TOWLDS - Tow Load Secondary

2) The Knuckle Assembly from Flight #8 was also used for Flight #9. A calibration check on cable angles LCAAGL and VCAAGL verified the calibrations did not change.

Tony Branco
Instrumentation Engineer

Document 38. Eclipse, EXD-01 Flight 9, 5 February 1998, Instrumentation Status Report, Tony Branco, Instrumentation Engineer

WATR Support
Eclipse
EXD-01 Flight #9
February 5, 1998

Mission Control Center:
NASA 1 (Blue Room), TRAPS 1, MFTS and MOF TM Site were used for this
flight. MOF TM Site tracked EXD-01 during takeoff and landing, and MFTS TM
site tracked the EXD-01 during flight. TV1 and TV3 provided ground support
for the EXD-01 and C-141A during take-off of flight 9. GPS data recorded on
EXD-01, and on C-141A.

Problems encountered:
During flight 9 the GPS receivers were not turned, on resulting in a loss
of GPS data for this flight. The problem was reduced in significance due to the
availability of GPS data from the DMA base station. However, the data from this source
is only updated once every 20 seconds.

Changes requested:
1) A new lineup was checked out for calibration changes in CIMS file for following
parameters : LCAAGL, VCAAGL, TOWLDP, TOWLDS

Debra Randall
Test Information Engineer / FE

Document 39. WATR [Western Area Test Range] Support, EXD-01 Flight 9, 5 February 1998,
Debra Randall, Test Information Engineer/FE

EXD-01 NASA 0130
Flight # 9

<u>QF-106 S/N 59-0130</u>

Flight Date: February 5, 1998

Pilot Time Logged: 1.5 hrs flying

Changes Since Flight #F8
- Structural post flight revealed no structural damage.
- No changes were made to the basic configuration of the aircraft.
- The Vectran rope was marked with cloth tape and paint every 100 feet using the technique used on the prior flight.
- The knuckle remains the same as on the last flight and a new frangible link was installed.
- A load cell was installed on the C-141 end of the tow rope between the three pin connector and the end of the rope. An additional end fitting was use with the rope and a special fitting was fabricated to fit the load cell and into the three pin connector. The load cell was wired to a laptop computer that was located at the video pallet in front of the load pallet. Mark Nunnelee was the technician that flew with the system and recorded the loads for the entire flight. A quick disconnect was use on the cable to the computer for rope jettison and the load cell went with the rope when dumped in the PIRA.
- Some procedural changes made to accommodate Wolfe Air tapping.

Flight #F9
Ground Operations:
- Ground operations were normal and were within 5 minutes of the schedule when both aircraft took the runway. Because rwy 22 will be used on this flight, the road between General Hill and Wolfe was be closed by the Security Police. It was scheduled to close at 0730 with a 0740 take-off planned. We stuck to our original schedule even though the take-off time was delayed until 0740. We figured that the additional delay time would be eaten up with the helicopter taping. It worked out well, and the flight held for just a short period of time before takeoff at 0735.
- Only minor problem when the helicopter flying around the operation got too close and blew sand and the rope around.
- One additional call was made by the control room during the takeoff operation. We made a 20 second call and had the helicopter acknowledge to ensure he was clearing the active runway.

Flight Operations:
- No control room issue noted
- Another fuel imbalance occurred during the flight and was worked by shutting off the appropriate boost pump.
- The Eclipse separated from the tow rope by breaking the frangible link.
- Knuckle remained intact with no damage.
- No post flight issues noted.
- A small amount of data was not gathered in the first part of the flight due to a system problem.

Mark Collard
Operations Engineer

Document 40. EXD-01 NASA 0130, Flight 9, Mark Collard, Operations Engineer

Eclipse
EXD-01 Flight 10
Sixth Tethered/Release Flight
February 6, 1998

Project Manager's Comments

The "Luck of the Eclipse" pulled us through the weather window again today. At first I had thought that Eclipse was just lucky when it came to weather, then I learned that our meteorologist has an artful touch to his forecasting ability. Thanks again, Casey Donohue. NO ONE other than Eclipse people thought we would actually have a weather window this morning! Bad day for rope photographs, but we did get our high altitude aero damping data!

It was a very challenging morning. I think if we had come across just one more obstacle Mark Collard would have thrown up his hands in total frustration. But he didn't. He patiently kept after it. And so did the rest of us.

I believe the reason we were able to handle today's effort is the same reason that we have had such an amazing series of successes: lots of thought, communication, and preparation. It's not magic, it's real.

I advise everyone to reflect on their Eclipse experience, take the personal lessons that you have learned, and apply it to your future endeavors. See if you can quantify them so that you can use them again and help improve project processes. Coach people, advise them on their practices, and clarify problem areas. I believe that we had some fundamental project practices that enabled everyone to work together effectively as a team.

A couple of my observations are:
1) Our various roles and responsibilities were clear and defined. This let individuals do their jobs and be responsible and accountable for their scope of effort. Even though we helped each other, we knew where one job left off and another started.

This enabled people to work extremely efficiently. An example is the Eclipse engineering team. Al Bowers "owned" the technical decisions on the project. He knew that neither I, the pilot, the crew chief, avionics lead, nor the ops engineer would make them for him. He also knew and respected the boundaries of his technical teams. He led them by orchestrating their decisions, and provided a climate that fostered communication and healthy debate. I felt that he was very effective.

Document 41. Project Manager's Comments, Eclipse, EXD-01 Flight 10, Sixth Tethered/ Release Flight, 6 February 1998, Carol A. Reukauf

2) Rigor. I think that this project team had a high degree of rigor in the way it interfaced. This extended from simply getting to a meeting on time (which on Eclipse was the norm), to making sure that all the elements were lined up for ground tests. Bill Lokos and Todd Peters were always prepared and thorough. We never had to think about it.

3) Listening. The Eclipse team listened to one another. This is tough for most projects. Many times individuals know their technical area well, and isolate their decisions from the influence of the rest of the team. By our being open to listening to the considerations of other team members, there were several significant decisions that were influenced by the input of a peripheral person's observations. How often did Jim Colombo Murray or Joe Gera summarize their issues very clearly and candidly? I, too, felt that some of the group forums seemed excessive, but in hindsight, I know that they were nearly always productive and value added.

So, today's mission concluded the flight test phase our very interesting little flight research project -- the one with the "very high cool factor," per Stephen Hoang at FRR. Now we have great data. And the fun continues for the research folks. And for KST and their future ambitions.

It's been a great experience!

Carol A. Reukauf

Eclipse F10 Flight Notes

Recently a comment was made that we towed in the rain on the final Eclipse tethered flight. While I can understand how it may be perceived that this was the case, I would like to set the record straight -- we did NOT tow in the rain or in IMC conditions at any time.

For those of you in the business of passing judgement on the safety of flight operations, I offer the following clarification.

Preflight Weather
The flight was scheduled for a forcasted temporary break in the weather. Conditions consisted of cloud layers at 6,000, 12,000, and 23,000 ft. Surface winds were southeast at ten kts. As anticipated, the weather impacted the flight plan (maneuvering to avoid clouds) and deteriorated during the flight.

Ground Operations
The weather was looking ominous (especially just northwest of Edwards) so I requested Gordo to take off early in the chase F-18 and give us a PIREP. He verified we could maintain VMC with sufficient ground reference throughout the climb to above 22,000 ft. As we progressed towards takeoff he was questioned by the control room and he restated we could do the planned profile to altitude due to the numerous breaks in the cloud cover and the higher ceilings to the east.

Flight Operations
We were able to easily maintain lateral and vertical spacing from the cloud layers during the climb to above 20K feet. The final portion of the climb to 25K did require some maneuvering and turning to maintain VMC as we spiralled up near some vertically developed clouds.

At altitude the nose-mounted video camera must have cold-soaked as the telemetered video got fuzzy. We were questioned by the control room regarding our flight conditions but I replied that we were VMC.

We performed most of the flight cards but did not complete all the points or do a couple of needed repeats because the weather was beginning to close up towards Edwards. Although the EXD still retained the same IFR capability as a stock QF-106, we had no intention of allowing the weather to close in beneath us since we wanted to maintain ground reference throughout the release and rope drop.

While there was rain between the flight and the airfield, we remained clear of clouds during our tethered descent to below the lowest broken layer. I chose to release and do a quick base key entry because Gordo was nearing bingo fuel and I wanted to allow all aircraft to land prior to the rain and gusty winds hitting. Although there was a small shower just southwest of the airfiled, I could maintain sufficient reference to perform a short right base key entry for a simulated flameout landing. I encountered my first raindrops during the landing. As I was taxiing into the NASA ramp the rain began in earnest.

At no time did the C-141 crew or I observe any precipitation during our tethered flight. I don't think Gordo did either but I don't want to speak for him.

Mark P. Stucky
Eclipse Project Pilot

Document 42. Eclipse F10 Flight Notes, Mark P. Stucky, Eclipse Project Pilot

Eclipse
Pilot's Flight Test Report
EXD-01 Flight 10 - 6th (Final) Tethered Flight, Feb. 6, 1998

Overview
This final tethered flight was to study the effects of altitude on the rope and tethered aircraft dynamics. The plan was to climb to the maximum altitude allowed within C-141A aircrew physiological constraints (25,000 ft.), C-141A performance limitations, or adequate tethered stability and control.

Flight above 18,000 ft. required the C-141A crew to accomplish the following prior to climbing above 10,000 ft.:
1. The orderly shutdown of the C-141A instrumentation pallet
2. Thirty minutes of pre-breathing 100% oxygen

Weather
The flight was scheduled for a forcasted break in the weather and luckily the weather cooperated. Conditions consisted of cloud layers at 6,000, 12,000, and 23,000 ft. Surface winds were southeast at ten kts. The meteorologist predicted the winds would switch westerly so the decision was made for a Runway 22 takeoff. The weather impacted the flight plan (maneuvering to avoid clouds) and deteriorated during the flight.

Ground Operations
We got to a late start because of some last minute instrumentation problems requiring a quick swap of the knuckle. Additional delays were caused by the combination of Runway 22 operations and deconfliction with other flight operations. The day-of-flight temperature was warmer than for earlier flights (45°), however engine start acceleration was still slow. The weather was looking ominous, so Gordo took off early in the chase F-18 and verified we could maintain VMC with sufficient ground reference.

Tethered Takeoff
The winds did not switch and were right at the crosswind/tailwind limit of ten knots. Takeoff was uneventful, and some turbulence was noted during the turnout reversal which manifested itself as lateral looseness.

Flight Cards
Climb Rate Effects on Stability - Clean (3K ft. - 10K ft.). Because of the turn reversals, only one data point was achieved prior to level off at 10,000 ft.

Stability Boundary Investigation - Clean. Once again, I was lower than the standard nominal position after level off. Point 2b (medium low) was really at the low position (so we skipped repeating it as a 2c point). The rope started some "leaf spring" oscillations, which required climbing and speed brakes to damp out. These leaf spring oscillations did not appear to effect the tension or stability and control, but would increase the post flight data reduction uncertainty because the rope angles during the photos will not exactly match the angles when the doublets were commenced several seconds later.

Climbing up at the 2e point (unstable high - but below the wake) an unstable dutch roll oscillation occurred which hit nearly 60° bank prior to initiating

Document 43. Eclipse Pilot's Flight Test Report, EXD-01 Flight 10—6th (Final) Tethered Flight, 6 February 1998, Mark P. Stucky, Eclipse Project Pilot

Eclipse
Pilot's Flight Test Report
EXD-01 Flight 10 - 6th (Final) Tethered Flight, Feb. 6, 1998

recovery. Because of the instability and the pitot boom / rope clearance issues, I chose not to do a lat/dir doublet at the 2f (above the wake) position. Holding fixed controls damped the oscillation immediately.

C-141A Control Input Effects. We chose not to do this card because the thirty minute timer for C-141A aircrew pre-breathing was just completing.

Climb Rate Effects on Stability - Clean (10K ft. - 25K ft.). Handling qualities remained good throughout the climb. The C-141A, which initially started the climb with the two outboard engines at idle, was at full climb power nearing 25K ft. and the climb rate had decreased to less than 1000 fpm. We were doing quite a bit of maneuvering to steer for holes in the cloud layers so the effective climb rate was even less. We leveled out maintaining VFR beneath a cloud layer which was near 25K ft.

Stability Boundary Investigation - Clean @ 25K'. In the medium low 2b position a rope leaf spring action of approximately +/- 15 ft. occurred. In the high (below the wake) 2e position I was forced to initiate a recovery due to lateral oscillations in the rope. This required the use of speed brakes to dampen. An unstable dutch roll also occurred after the lat/dir doublet.

I was at this high unstable position when the C-141A started a turn. This forced a quick descent to maintain control. **The tow aircraft should request permission or give a warning call prior to maneuvering at unstable flight conditions.** We were running out of time (due to chase fuel) as well as weather so point 2f was not done.

Trim in Turns. The EXD did not appear to track as nicely in the clean configuration at altitude. During steeper turns the AOA was approaching "minimum safe" and there was light to moderate aerodynamic buffet. Conditions were worse in the medium high (straight rope) position, and pilot workload was moderate due to looseness about all axis.

Descent. We had wanted to do a manual release at altitude over the PIRA but were constrained by the weather, so we chose to remain on tow for the descent to VMC beneath the broken layers. The C-141A descended at nearly 2000 fpm (the highest rate of descent for the program) which required keeping the speedbrakes deployed to maintain rope tension and system stability.

Manual Release. Tension values were approximately 3000 lb. (we were still in a descent) so I lowered the landing gear and kept the speed brakes deployed. With a tension value above 5,000 lb., I pulled the manual release handle at approximately 9,000 ft. and 190 KIAS. I maneuvered to the right and observed the knuckle immediately separate from the tow rope on the first rebound.

Eclipse
Pilot's Flight Test Report
EXD-01 Flight 10 - 6th (Final) Tethered Flight, Feb. 6, 1998

I cleaned up the aircraft and turned to enter a right base key position for a simulated flameout landing. I was low on energy which dictated a short and tight pattern. Landing was uneventful and the rain began as I taxied clear of the runway.

Mark P. Stucky
Eclipse Project Pilot

DAILY/INITIAL FLIGHT TEST REPORT		1. AIRCRAFT TYPE	2. SERIAL NUMBER
		C-141A	61-2775

3. CONDITIONS RELATIVE TO TEST

A. PROJECT / MISSION NO	B. FLIGHT NO / DATA POINTS	C. DATE
Eclipse	F-10, 6th/Final Towed Flight	06 FEB 98
D. LEFT SEAT *(Front Cockpit)*	**E. FUEL LOAD**	**F. JON**
Capt Stu Farmer	38,000	C9703900
G. RIGHT SEAT *(Rear Cockpit)*	**H. START UP GR WT / CG**	**I. WEATHER**
Maj Kelly Latimer	206,000 lbs / 28.8%	70 SCT 120 BKN 200 OVC, LT Rain
J. TO TIME / SORTIE TIME	**K. CONFIGURATION / LOADING, SOFTWARE**	**L. SURFACE CONDITIONS**
1601 Z / 1.6 hrs	Petal Doors Removed / Tow Config	48°F/ Winds 070 @ 10kts
M. CHASE ACFT / SERIAL NO	**N. CHASE CREW**	**O. CHASE TO TIME / SORTIE TIME**
F-18 NASA 852		

4. PURPOSE OF FLIGHT / TEST POINTS

This flight was the sixth and final towed flight of the Eclipse program. The all Vectran rope was again used and the tow was terminated by a normal release of the EXD-01. Planned for this flight was a towed climb to 25,000' MSL and release at that altitude. Test cards consisted of climb rate effects, stability boundary investigation, and trim in turns. The C-141 control inputs card was planned but not completed.

5. RESULTS OF TESTS *(Continue on reverse if needed)*

There was a slight hold prior to engine start as a knuckle hardware issue was worked by the EXD-01. The C-141 made a normal engine start at 1441 Z and began taxi to RWY 22 at 1447 Z. The C-141 arrived at the last chance area for RWY 22 at 1454 Z. The EXD-01 was just nearing engine start as the C-141 arrived. There was an extended delay prior to taking the runway while holding for a C-130J operation. Weather was marginal due to cloud cover, but chase took off and reported a climb in the clear could be accomplished.

EDW weather observation for this time was winds 100°@11 kts, temperature 48°F, clouds few 5,000', scattered 7,000', broken 12,000' and overcast 20,000'. Actual weather was better than this reported weather, but there were numerous thin broken clouds in the area.

The C-141 taxied onto the runway at 1532 Z. The tailwind caused a high concentration of C-141 engine exhaust fumes to enter the rear of the C-141 and the C-141 deployed the thrust reversers early to minimize this. At 1540 Z the rope was aboard the C-141. Slack was removed at 1549 Z. Initially, the loadmaster called slack removed but slack remained at the EXD-01 end of the rope and the C-141 continued creeping forward until the EXD-01 called slack removed.

The C-141 crew began breathing 100% oxygen as required for unpressurized flight above 18,000' MSL. The communications setup for the loadmaster, as in previous flights, required him to remain on hot mike for interphone communications as his push-to-talk switch on the extended comm cord was used for transmitting on mission frequency. The breathing of the loadmaster on hot mike was a significant distracter for all communications throughout the flight.

There was a hold for several minutes waiting for the C-130J to clear the runway and for the new takeoff time of 1600 Z for Lancaster Boulevard to be closed. Tension was set at 1558 Z. The flight was ready for takeoff at 1600 Z. Chase made the airborne pickup calls and brake release was at 1600:55 Z. C-141 fuel at takeoff was approximately 32,000 lbs. Towed takeoff was normal.

------ CONTINUES NEXT PAGE ------

6. RECOMMENDATIONS

Aircraft tow can be safely conducted and with experience could be treated as a normal operation.

Use of hot mike should be avoided by personnel on oxygen.

COMPLETED BY	SIGNATURE	DATE
Morgan LaVake, Test Conductor		06 FEB 98

AFSC Form 5314 NOV 86 REPLACES AFFTC FORM 305 MAR 84 WHICH WILL BE USED

Document 44. Daily/Initial Flight Test Report, C-141A, 61-2775, 06 Feb. 98, Morgan LaVake, Test Conductor

Section 5 continued.......

Climbout was at 190 KIAS and 2000 fpm rate of climb. Doublets were performed in the climb by the EXD-01. Several turns were made during the climb for weather avoidance. There was light turbulence for much of the climb. The flight leveled at 10,000' MSL in light chop at about 1605 Z. Cruise EPR setting was approximately 1.3. The EXD-01 began the Stability Boundary Investigation - Clean test card. Turbulence increased somewhat beyond the previous light chop. During one test point, the EXD-01 deployed speedbrakes to recover the stable tow position. Winds aloft as reported by chase were 200° @ 40 kts at 1614 Z at 10,000 MSL. At 1617 Z the flight entered smooth air. As seen on previous flights, the EXD-01 doublets in the higher tow positions caused slight pitch oscillations in the C-141. Attempting to maintain the high tow position above the C-141 wake, the EXD-01 had several reported encounters with the C-141 vortices that moved him into a lower tow position.

The NASA control room decided to delete the C-141 Control Inputs test card for the purposes of time and fuel management. At approximately 1625 Z the TPS data pallet was shut down and oxygen pre-breathe requirements had been met. The flight then began climbing to higher altitude.

Passing through approximately 11,500' the TPS data pallet barometric switch activated and removed all power from the pallet.

In the climb, power was increased with altitude to maintain the 2000 fpm rate of climb. Above approximately 20,000' the 2000 fpm climb could not be maintained and climb rate decreased to about 1500 fpm with power set at N$_{RT}$. At several times during the climb, doublets were performed for data by the EXD-01. Clouds in the area necessitated moderate maneuvering for avoidance. At 1635 Z the flight leveled at 23,800' MSL to remain below the overcast. C-141 fuel was 21,000 lbs.

Testing proceeded with the Stability Boundary Investigation - Clean and the Trim In Turns test cards. At altitude, C-141 power was approximately 1.4 EPR on all four engines to maintain level flight.

Because of cloud cover, the flight began a descent to a lower altitude for the planned tow release. The EXD-01 released at 7,300' MSL at 1714:45 Z while the flight was in a 500-1000 fpm descent at 190 KIAS. Release was by planned activation of the manual release by the EXD-01. The knuckle fitting was reported to separate from the rope almost immediately after release.

The C-141 dropped the rope into the PB8 drop zone. The primary and secondary guillotines were activated in close succession, and the rope departed the aircraft upon activation of the secondary release. The primary guillotine did not function at all. The blade remained in the cocked position and was found to be jammed in that position. Application of full force to the activation tether would not move the release catch holding the blade in the cocked position.

The C-141 returned to EDW for landing. Weather continued to deteriorate and the C-141 approach and landing were in light to moderate rain. The C-141 landed on Rwy 04 at 1729 Z with approximately 12,000 lbs of fuel remaining.

The test pallet tapes were provided to NASA for processing. The NASA Ashtec GPS was removed from the aircraft and returned to NASA for downloading.

Aerodynamics
Flight Report
Flight 10 (6th towed flight)
February 6, 1998

Flight Crew:
EXD-01 (NASA QF-106A 59-0130): Mark "Forger" Stucky
C-141A (USAF C-141A 61-7775):   Stu Farmer, Kelly Latimer, Morgan LeVake,
                                 John Stahl, Dana Brink, Ken Drucker

The last Eclipse mission was flown on Friday, February 6, after a rapid
turnaround by the aircraft crew. This was an altitude mission, the
objective was to exceed 20k ft msl (weather permitting). The official
forecast was ominous; but local weather predictions predicted there
would be a break in the morning long enough to fly the mission.
Because of the altitude requirement and the unpressurized C-141A, a 30
minute prebreathe was required of all C-141A flight crew before
exceeding 10k ft msl. Also to monitor the crew, phys-techs
(physiology technicians) were required onboard for the entire
mission.

On this flight, a 10 knot tailwind component and a 7 knot
crosswind component was encountered. Chase reported promising
conditions, a haze layer between 15 and 18 k, but then clear air up to
a ragged ceiling at about 24k ft msl.

Takeoff was performed followed by some maneuvers at 10 k ft msl while
waiting for the 1/2 hour C-141A crew pre-breathe. After that, the flight
continued upstairs to 23k ft msl. Because of limited test time, only stability
boundaries and doublets could be performed.

The rope was marked on this day with tape and paint, but cloud cover
prevented effective photography of the rope sail. Holes in the clouds were
noted through to blue sky above, but limited use of these holes could be made
because of maneuvering limitations.

The EXD-01 pilot noted increased workload to maintain tow
position (the pilot asked the mission controller to read the cards to
him on the radio, he was unable to spend time reading the cards on his
own; a clear indication of increased workload).

It was hoped that the C-141A could tow the EXD-01 to high key position and
the EXD-01 could execute a release and landing without touching the
throttles.

Document 45. Aerodynamics, Eclipse EXD-01 Flight 10, 6 February 1998, Al Bowers, Chief
Engineer

The rain and clouds prevented such a plan from occurring, both aircraft descended to 8500 ft msl (to remain clear of clouds), and the EXD-01 used the manual release (thus demonstrating the last of the three release options; electrical release, manual release, and frangible link break). The C-141A dropped the rope, and the two aircraft landed on the main runway just as the rain started to fall heavily.

Flight 10 (sixth towed flight)
Date: 02/06/98
Take off Time: 08 01 53
Release Time: 09 14 51
Landing Time: 09 17 33
Flight Time: 01 15 40
Tow Time: 01 13 58
Total Tow Time: 05 34 29

Maximum Mach: 0.501 (flight 10, 6th towed flight)
Maximum Altitude: 24,684 ft msl (flight 10, 6th towed flight)

Al Bowers
Chief Engineer

Eclipse
Flight Controls
EXD-01 Flight #10
February 6, 1998

Brake release and takeoff.

Brake release was very smooth, resulting in less than 0.25 g's and less than 11,000 lbs of peak tow rope tension. During the relatively steep initial climb by the tow plane a peak tow rope tension of 12,000 lbs was observed.

Climb to 10,000 feet.

The climb to altitude was made at 2,000 ft/min. The only difference from previous flight was the presence of turbulence or 'light chop' that was reflected mainly on roll rate response of the EXD-01.

Stability boundary tests.

During this flight the stability boundaries were investigated only in the clean configuration at two test altitudes: 10,000 and 24,500 feet.

Just before moving to the lowest tow position at the 10,000 ft test altitude, the pilot reported light chop; at the same time the lower gimbal limit was reached and large oscillations of the tow rope were observed. The pilot used the speed brakes to damp out the oscillations. On one occasion a brief throttle input was made in an attempt to attenuate the tow rope tension as a large slack was being taken up. The high tow positions just below the wake were characterized by the pilot as requiring high workload due to the extremely dynamic behavior of the tow rope. In contrast, both the workload and the dynamics of the tow rope were relatively mild above the wake. At the 24,500-ft test altitude the EXD-01 airplane felt generally looser to the pilot in spite of the smoother air at the higher altitude.

Varying the low tow positions had relatively small effect on the response of the airplane. Even in the lowest tow position the airplane was stable both longitudinally and lateral-directionally. In the high-unstable position the tow rope developed extremely complex motions, but the tow rope tension remained at moderate values. The large amplitude motion of the tow rope was damped by the pilot by modulating the speed brakes. No maneuvers above the wake were made at the higher test altitude.

Turning flight.

Trimmed flight in turns was evaluated at 24,500 feet in the clean configuration. The EXD-01 did not track the tow plane as well as at the lower altitude and, according to the pilot, "it had to be hand-flown continuously as a conventional sailplane."

Descents on tow.

Although there were no specific test points for rates of descent scheduled for this flight, a long descent was made from 24,500 feet to 7,500 for the frangible link separation. The descent was made with the speed brakes open at approximately 2,000 ft/min both in smooth air, and in light turbulence. No tendency for tow rope slack was noted at any time during the descent.

Tow rope release.
A normal tow rope release, using the pneu-mechanical release system, was made at 7,500 feet over the PIRA. The pilot reported that release hardware separated from the tow rope almost immediately.

Joe Gera

Flight Mechanics
Eclipse
EXD-01 Flight #10
February 6, 1998

Initial reports from the chase photographer indicate that lighting
conditions were poor for documenting rope sail. No useable photo data
is expected from this flight.

Jim Murray

Document 47. Flight Mechanics, Eclipse, EXD-01 Flight 10, 6 February 1998, Jim Murray

Structures Report
Eclipse
EXD-01 Flight #10
February 6, 1998

This flight was the only one to exceed 10,000 feet altitude - going to about
25,000 feet. Additionally, it overcame the most challenging weather conditions
of the flight program threading its way in limited clear air amongst considerable
clouds and finally landing in rain after a full flight.

System configuration was similar to last flight with the exception of having no
load cell assembly at the C-141A. The rope was 1000 feet of continuous
Vectran with the now routine painted tape marker segments.

Frangible link load signal zero load offsets were small. The takeoff roll tow load
profile was generally similar to previous cases and produced a moderate
maximum magnitude of about 12,000 pounds at rotation and a first cycle peak
of 10,500 pounds. No slack occurred during takeoff. This was perhaps the
smoothest looking (from a loads standpoint) takeoff of the flight program.

Peak maneuver loads of about 12,000 pounds occurred during a
lateral/directional doublet at 10,000 feet MSL, 190 KCAS, clean. This condition
produced a very dynamic rope response oscillating between the peak load and
zero load (slack) and was manually stopped by pilot recovery input. A second
peak maneuver load occurred at about 25,000 MSL, 190 KCAS, clean during a
longitudinal doublet performed at high tow (above wake). Here again the loads
oscillated between up to 12,000 pounds and zero load (slack) , was very
dynamic and visually impressive and was manually stopped by pilot recovery
input. These maneuvers appeared to be able to produce unacceptably high
loads or excess slack.

Rope release was accomplished by pulling the manual release handle at about
190 KIAS, 7,000 feet alt, with a rope tension of 4,500 pounds. Upon release the
rope and end assembly sprang well forward and then rebounded straight back
with the end assembly apparently separating at the frangible link during the
beginning of the first whip cycle. The end assembly parts then traveled aft past
one side of the EXD-01 proving the uncertainty of post-link-failure kinematics
and showing the value of the pilot's procedure of developing lateral offset prior
to release. Of all the attention paid to the many details, this too was worth the
trouble.

Bill Lokos

Document 48. Structures Report, EXD-01 Flight 10, 6 February 1998, Bill Lokos

ECLIPSE
EXD-01 Flight 10
Weather Summary
February 6, 1998

Strong mid-pacific storms that have been moving into the edwards area were a concern leading the last Eclipse flight on February 6. The forecast from the Air Force had rain in the area by sunrise with winds gusting out of the south at 30 knots. The forecast was unfavorable for flight. However, after looking at the data it appeared that their would be an opportunity to fly, as some of the moisture from the system would be held back by the mountains until the cold front, and associated precipitation, moved closer to the area. The forecast was amended to indicate broken skies at 20,000 feet and scattered to broken clouds at 15,000 feet with a few clouds at 5,000 feet. The winds were still forecast to blow from the south but at speeds near 10 to 15 knots. Precipitation was forecast to stay in the mountain areas through mid-morning. Turbulence was forecast to be light to moderate from the surface to 12,000 feet.

Flight day weather observations in the early morning were close to operational limits. The primary concern was the winds. Wind tower observations before 05:00 PST indicated the winds were northeasterly at 10-12 knots. Forecasts from the Air Force indicated the winds would turn south by 06:00 PST with gust to 25 knots. The forecast also included rain by 08:00 PST with visibility reduced to less than 5 miles. Due to the wind forecast and observations from nearby stations, it was decided to use runway 22 to ensure that the mission will not violate the >10 knot cross/tail wind. As flight time approached it was clear that the winds were not going to switch from the south. By takeoff time, 08:02 PST, the winds were northeasterly at 10 knots. Runway 22 crosswinds were near 7 knots with a tailwind component of 10 knots. The temperatures throughout the morning remained in the low 40's. The wind and temperature data were observed from wind towers located along the main runway. Tower 044 is located 4000 feet down the threshold of runway 04 and Tower 224 is 4000 feet down runway 22. Both towers measure data 30 feet above ground level. Sky conditions were observed to be broken altocumulus at 10,000 feet and scattered cumulus at 5,000 feet. Light turbulence was experienced between surface and 8,000 feet. Rain began to fall just after 10:00 PST as both vehicles were returning to base. Although the pilots in the C-141A had to fly around clouds and rain showers during the flight, the weather was acceptable for the final flight of the Eclipse program.

Casey Donohue

Document 49. Eclipse, EXD-01 Flight 10, Weather Summary, 6 February 1998, Casey Donohue

Eclipse
EXD-01 Flight 10
February 6, 1998
Instrumentation Status Report

1) New calibrations for this flight:

      TOWLDP - Tow Load Primary
      TOWLDS - Tow Load Secondary

2) Originally the Knuckle Assembly from Flight #7 was to be used for Flight #10. Upon examination of the Knuckle Assembly on the day of flight, it was discovered that the Vertical CPT cabling was damaged. A decision was made to perform a quick turn around of Knuckle Assemblies. Link #9 from the damaged assembly was installed on the knuckle assembly used in flight #8 and *9. Comparing vertical and lateral cable angle calibrations revealled a maximum deveation of 1.75 deg. The differences were noted with no saftey-of-flight or mission success concerns. A post flight calibration check will be performed.

Tony Branco
Instrumentation Engineer

Document 50. Eclipse, EXD-01 Flight 10, 6 February 1998, Instrumentation Status Report, Tony Branco, Instrumentation Engineer

WATR Support
Eclipse
EXD-01 Flight #10
February 6, 1998

Mission Control Center:
NASA 1 (Blue Room), TRAPS 1, MFTS, and MOF were used for this flight. MOF
TM Site tracked EXD-01 during takeoff and landing, and MFTS TM site tracked
the EXD-01 during flight. TV1 and TV3 provided ground support for the EXD-01
and C-141A during take-off of flight 10. GPS data recorded at 4800, on
EXD-01, and on C-141A.

Problems encountered:
During flight 10 the discovery was made that MOSES was incorrectly shutdown on
February 5, 1998 after the SRA flight. This required that some of Indigo's running PAGE
displays had to be rebooted. However, this did not affect the acquisition of Eclipse flight
data or signficantly interrupt flight 10.

Changes requested:
1) A new lineup was checked out for calibration changes in CIMS file for following
parameters : LCAAGL, VCAAGL, TOWLDP, TOWLDS

Debra Randall
Test Information Engineer / FE

Document 51. WATR [Western Area Test Range] Support, EXD-01 Flight 10, 6 February 1998,
Debra Randall, Test Information Engineer/FE

EXD-01 NASA 0130
Flight #10

<u>QF-106 S/N 59-0130</u>

Flight Date: **February 6, 1998**

Pilot Time Logged: 1.3 hrs flying

Changes Since Flight #F9
- Structural post flight revealed no structural damage.
- No changes were made to the basic configuration of the aircraft.
- The Vectran rope was marked with cloth tape and paint every 100 feet using the technique used on the prior flight.
- The knuckle remains the same as on the last flight and a new frangible link was installed.
- The electro-pneumatic release circuit breaker was pulled and collared.

Flight #F10
Ground Operations:
- During installation of the knuckle, one of the CPTs broke and the entire knuckle was changed out in 45 minutes and the DOF checks were delayed by 25 minutes. The crew did an outstanding job to pull it all together and minimize the delay.
- An additional delay was encountered when we had agreed to allow the C-130J aircraft on the runway to do a taxi test prior to our take-off. They seemed to be taxiing very slowly, and took forever to get on the runway. After all that wait, they did not perform the test due to unfavorable winds.
- By the time we were close to taking runway 22 (again for the predicted winds) and and it was nearing the time for the road to be closed, somehow the Security police got the word from some other person to close the road at 0730. When I called at 0745 to inform them we would like the road closed starting at 0750, they informed me it was already closed.
- Just prior to take-off, range control called and informed us that we had a conflict in the PIRA from 0800 to 0900 and that the F-16 had the priority. We informed the range that we would not enter the PIRA until after 0900 but would have to transit the PIRA after takeoff to climb into the complex.
- To top off all our other problems for the day, the weather was bad to the West and was looking worse all the time. We were getting a lot of questions about the weather from the Director of Operations. Chase was reporting clear to the East and the front seemed to be stationary over California City. Takeoff finally occurred at 0802.

Flight Operations:
- No control room issue noted.
- After the 30 minutes of 100% Oxygen pre-breathing, the flight was cleared to climb above 10,000 feet. The Instrumentation was shutdown also prior to starting the climb, as briefed.
- The flight crews had to be creative to find holes in the clouds while climbing.
- The Eclipse separated from the tow rope using the normal tow release of opening the jaws. However, the jaws were opened using the manual release Tee handle.
- Post flight inspection revealed no structural damage.

Mark Collard
Operations Engineer

Document 52. EXD-01 NASA 0130, Flight 10, Mark Collard, Operations Engineer

Eclipse
Acronyms and Definitions

AGL	Above ground level
AOA	Angle of attack
ARRIS	C-141A call sign
Clean	Low drag configruation, gear and speed brakes retracted
Dirty	High drag configuration, gear down, speed brakes extended
DOF	Day of Flight
EOR	End of Runway
EXD	EXD-01, shortened. Modified QF-106
FOD	Foreign Object Damage
GPS	Global Positioning System
HQR	Handling Qualities rating
KIAS	Knots indicated airspeed
MCC	Mission Control Center (control room)
MOF	Mobile Operations Facility
MSL	Mean seal level
PIRA	Precision Impact Range Area
RTB	Return to base
SFO	Simulated Flame Out
VMC	Visible Meteorological Conditions
VFR	Visble Flight Rules

Document 53. Eclipse Acronyms and Definitions

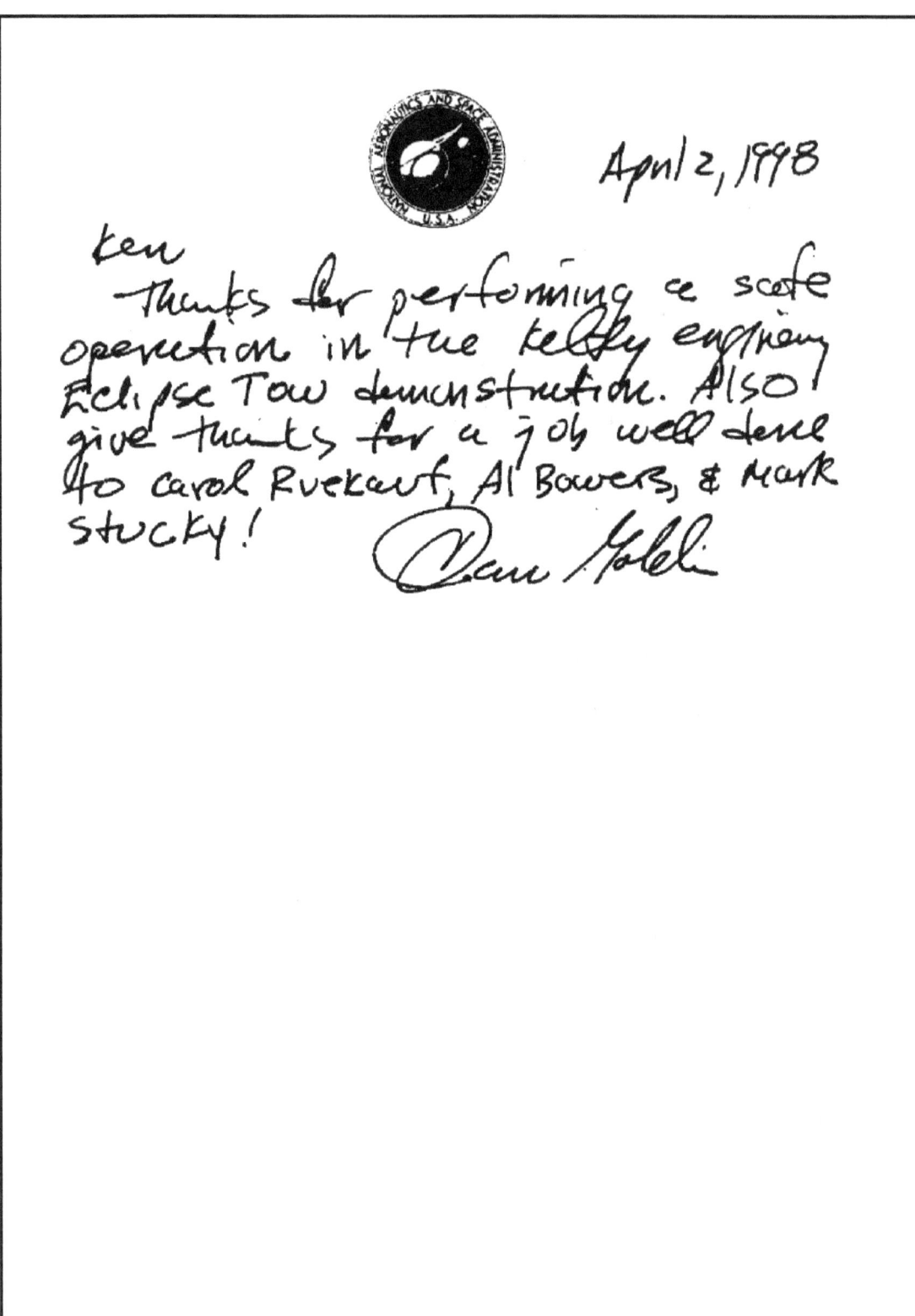

April 2, 1998

Ken
Thanks for performing a safe operation in the telly engineering Eclipse Tow demonstration. Also give thanks for a job well done to Carol Reukauf, Al Bowers, & Mark Stucky!

Dan Goldin

Document 54. Note, Dan Goldin to Ken [Szalai], 2 April 1998

The Eclipse Project

Document 55. Eclipse Project Pilot Mark P. Stucky's slides used at briefings

Background

Soaring Experience
- » Hang gliding since 1974
- » Sailplanes since 1981
- » Paragliders since 1992

Military pilot (USMC) 1981 – 1993

NASA – JSC 1993 – 1996
- » T–38
- » Shuttle Training Aircraft

NASA – DFRC 1996–1999
- » F–18 High AOA Research
- » F–16XL
- » Eclipse

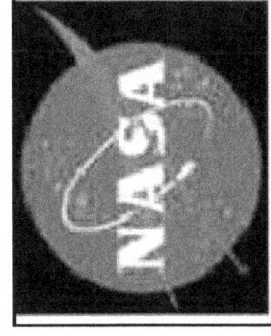

Eclipse Astroliner Concept

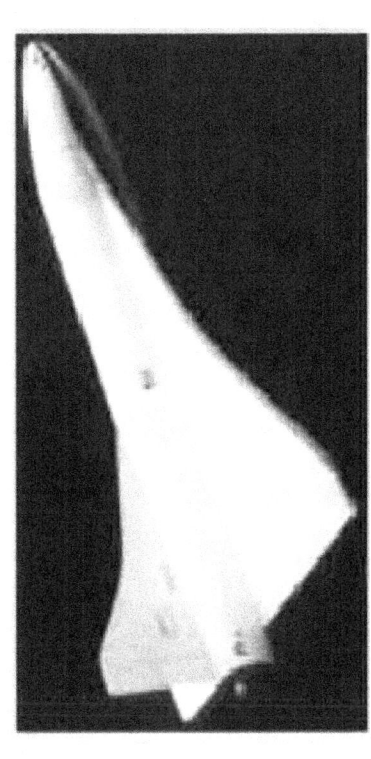

- Kelly Space & Technology patented a unique reusable space launch concept
- "Booster" stage is an aerotow to the stratosphere
- Astroliner ignites rocket engine, verifies systems, releases tow line, and climbs toward space
- Upper stage with payload separates at 400,000 ft / Mach 15
- Astroliner peaks at 600,000 ft then reenters for a runway landing

Historical
Precedents

Aircraft	Weight	Wingloading	Comments
Me–163B	9,500	45	German WWII rocket plane
He–280	7,100	31	German WWII jet
P–51B experiment	7,100	31	NACA Ames L/D
Mx–334	3,000	12	Northrup flying wing
I Ae 37	2,600	5	Highly swept delta wing
T–33	15,000	64	USAF "Longline" project
Sailplane	1000	8	Typical glider
EXD–01	30,000	43	Eclipse Demonstration
Astroliner	320,000	80	Proposed Eclipse RLV

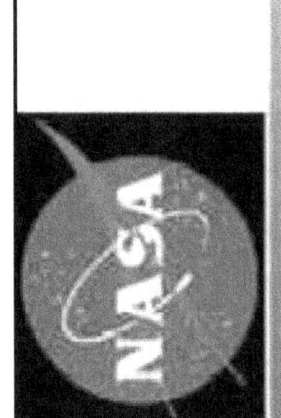

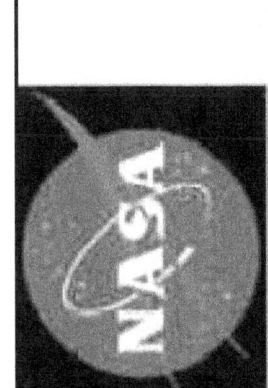

Program Philosophy

- NASA had been challenged to perform "Faster, Better, & Cheaper"

  » Eclipse Experimental Demonstrator Program was to be a low–cost proof–of–concept

    – Use existing hardware where practical

    – Design for a load factor of 225% limit load

    – Avoid gold–plating

Obvious Issues

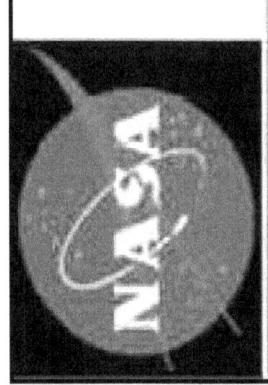

- Tow rope length and material
- Tow position
- Wake turbulence / tip vortices
- What strength weak link?
- Emergency Procedures
- Communication / Terminology

Pilot Training

- EXD Pilot
  - » USAF QF–106 checkout
  - » Proficiency flying
    - – Formation flight tasks at various slow speeds
    - – Low altitude rope break / waveoff practice
  - » Glider tow training
    - – Conventional sailplanes
    - – SWIFT 3–axis controlled hang glider
- C–141 Pilots
  - » Takeoff procedures & technique
- All pilots
  - » Integrated simulation

Eclipse Simulation

- Growth from 1-D batch to 6-D linked sims
  - » Fully Linked EXD, rope, & C-141 sim
- Varying degrees of "flyability"
- Used for integrated CRM / EP training
- Simulation became a research objective
- Simulation changed our procedures
  - » EXD takeoff configuration
  - » Tow position
- Simulation limitations
  - » Switchology
  - » Hook-up, tension, and brake release

EXD Cockpit

- Removed unnecessary hardware
- Manual release handle
- Tension displays
  - » Digital readout
  - » Color light bar
- HOTAS release mechanism
- Instrumentation control panel

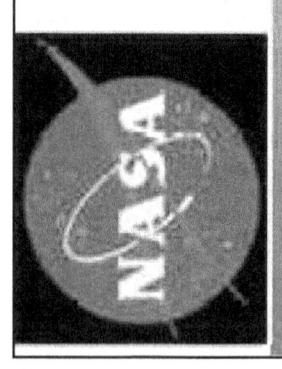

Attachment Hardware

- B–52 Drag Chute Release
  - » Reduced design effort
- Release Modifications
  - » Measure tow angles
  - » Releases only under load
  - » Electro–pneumatic actuation
  - » Manual backup
- Frangible (weak) Link
  - » Design limit of 24,000 +/- 200 lb.
  - » Dual channel load cell

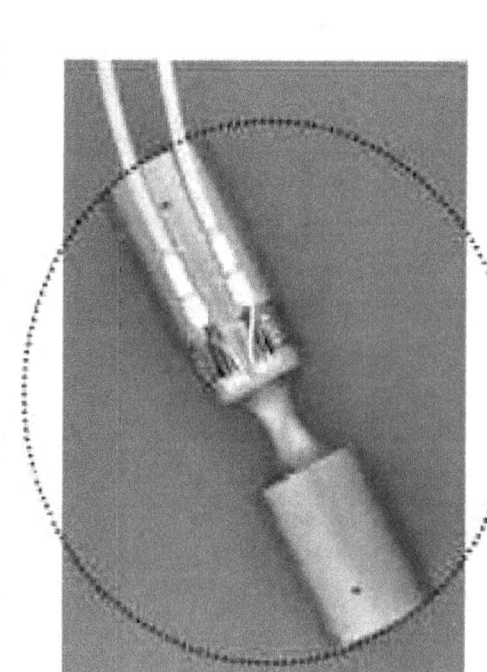

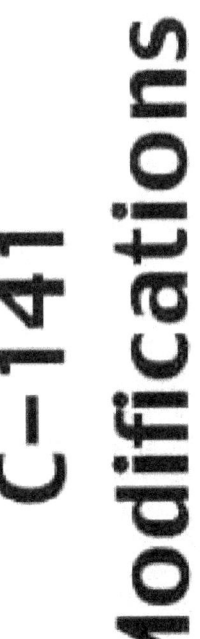
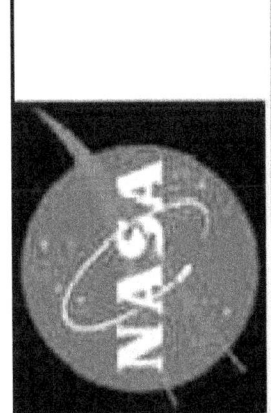

C-141
Modifications

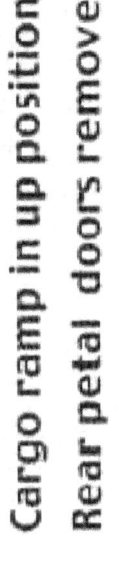

- Cargo ramp in up position
- Rear petal doors removed
  - » 200 KIAS limit
  - » Unpressurized cabin
- Instrumentation pallet
- NASA-supplied DGPS
- Rearward-looking video cam

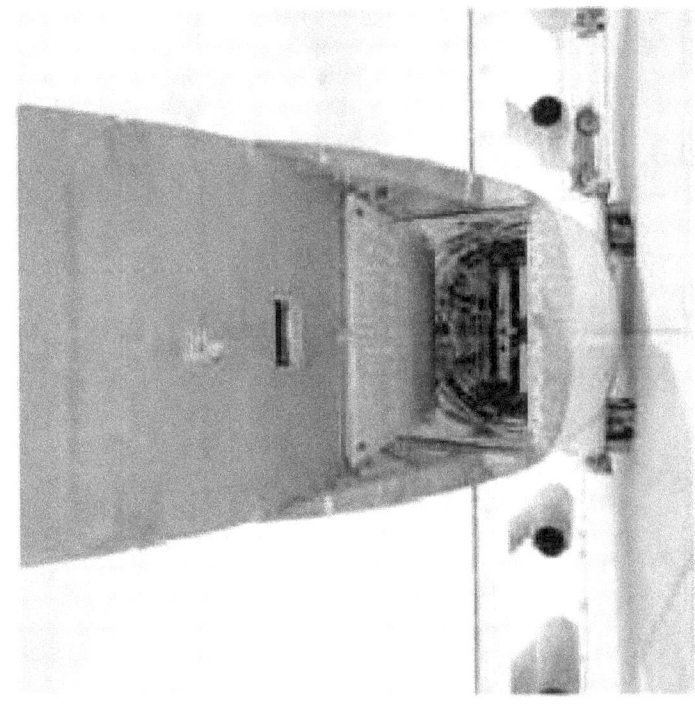

Tow Rope

- 1000 ft of 3/4" diameter Vectran
  - » 60,000 lb. tensile strength
  - » Low elongation
- Middle 50 ft – 8–ply nylon
  - » Acted as shock absorber
  - » Reduced pilot workload

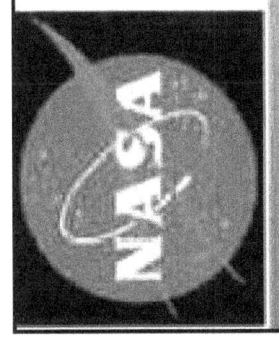

Untethered Tests – C-141 Takeoff

- Wake Vortice Investigation
  - » Separation during takeoff/climbout
- Takeoff Technique Analysis
- Difference from Flight Manual
  - » Nonstandard – must be repeatable
- Takeoff Performance
  - » Acceleration vs. Tension
  - » De-rated thrust to simulate tow

C–141 Takeoff Profile

- Wake Turbulence Separation
- Liftoff in the Middle of Stability Envelope
- Takeoff Profile for low tow
  - » Smooth trajectory important for minimizing bungee oscillations

$VVI = 1000$

$V_{T/O} \frac{fpm}{EXD} = 170$ KIAS

$H_{C-141} = 350$ ft AGL

$H_{C-141} = 200$ ft AGL

Vr_{C-141}

115 KIAS

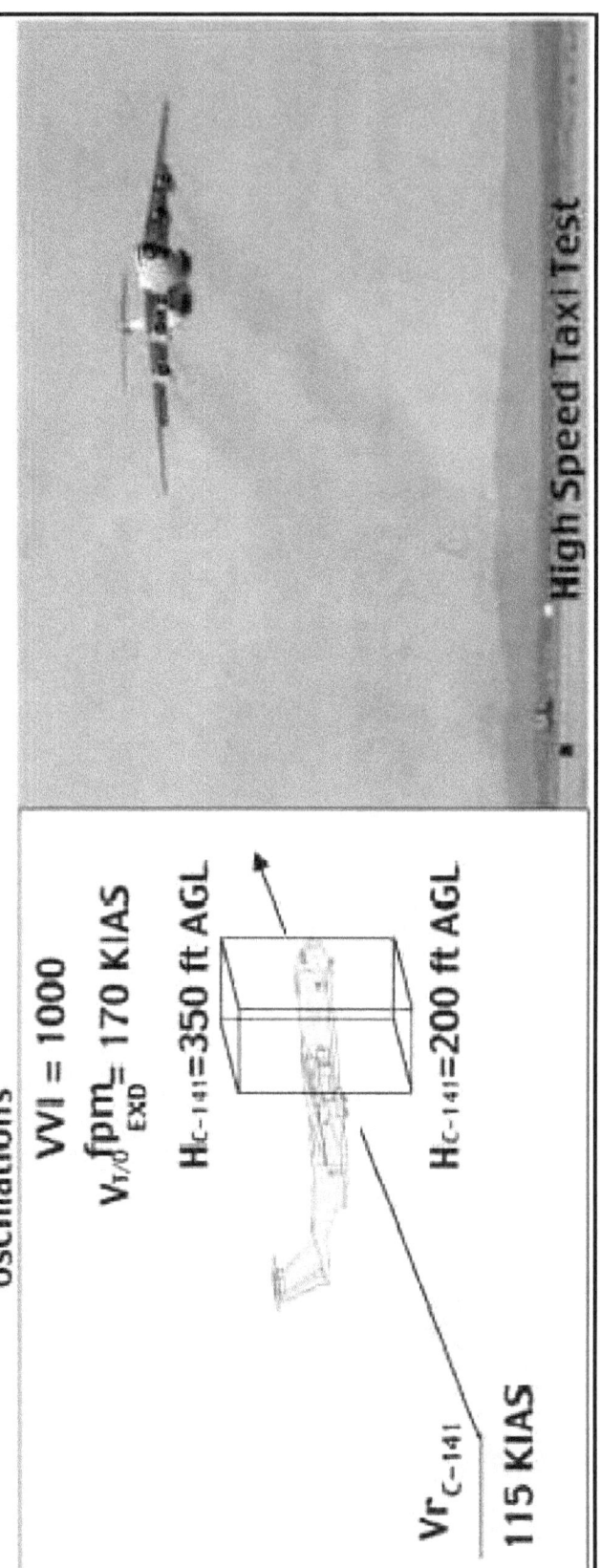

High Speed Taxi Test

Tethered Research Objectives

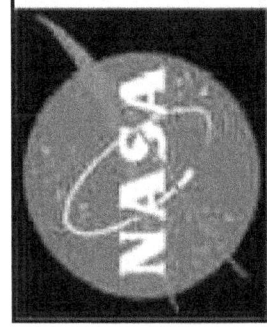

- Demonstrate the proof-of-concept

- Parameter estimation of EXD and tow rope

- Quantify tension changes due to configuration changes

- Investigate the controllability and handling qualities on tow (including climbs, descents, & maneuvering flight)

- Quantify trim changes due to rope sail

- Determine rope sail effect on tow rope damping

- Determine altitude effects on stability and control

Initial Results

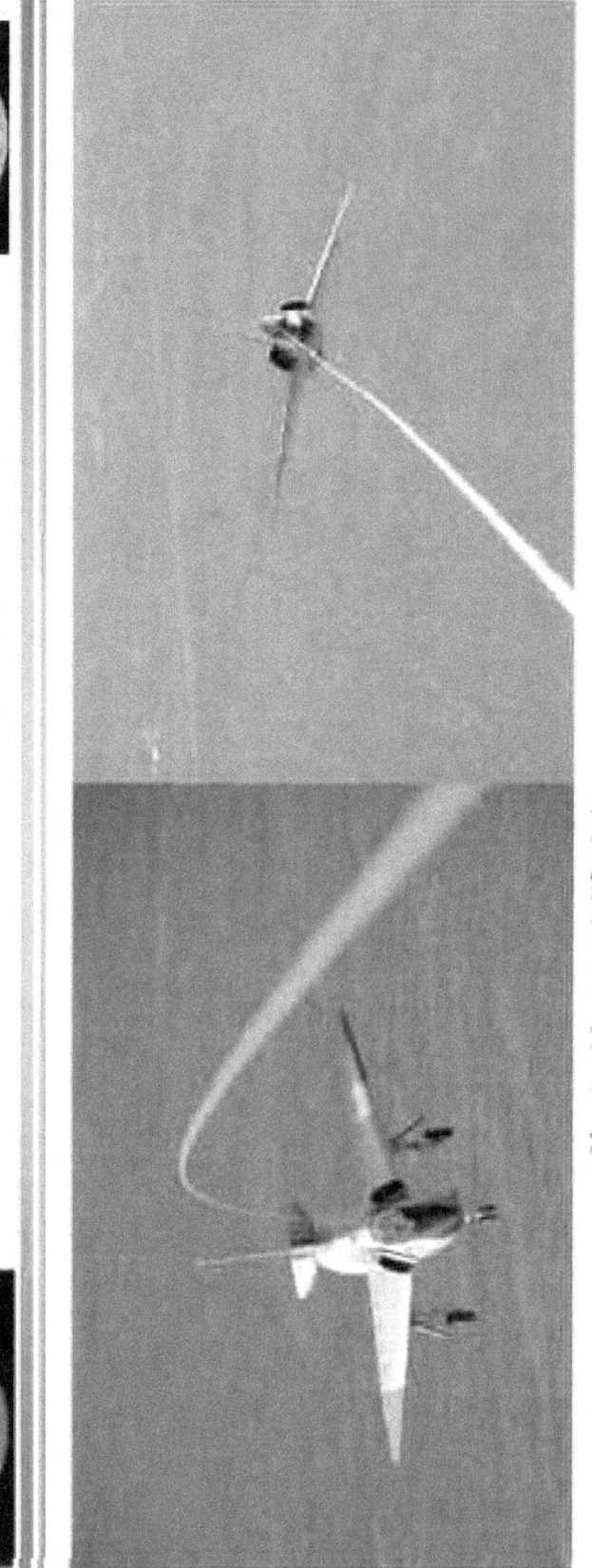

- Six tethered flights
- Tow rope aero forces increase system damping
- Predicted tow rope loads validated
- Large stable tow envelope
- Tow concept proven
- Research sim model being generated

Weak Link Release Procedure

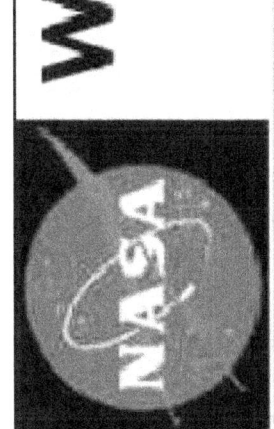

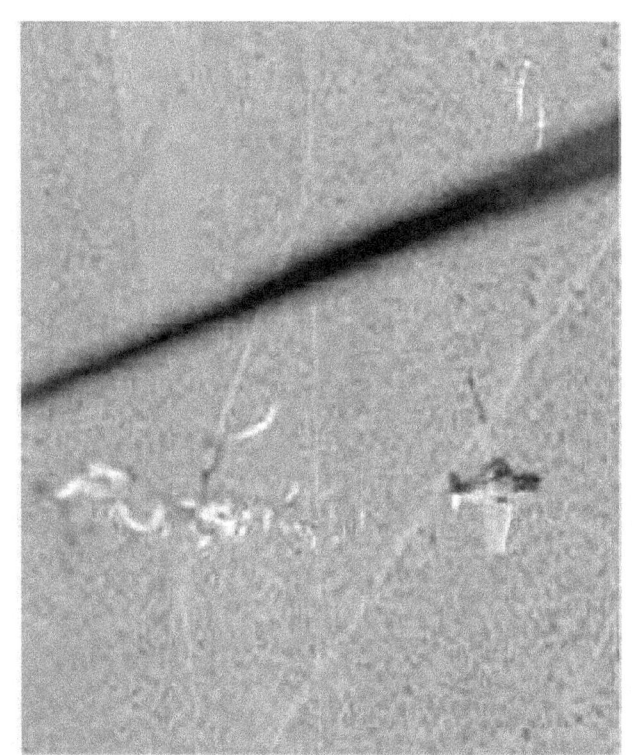

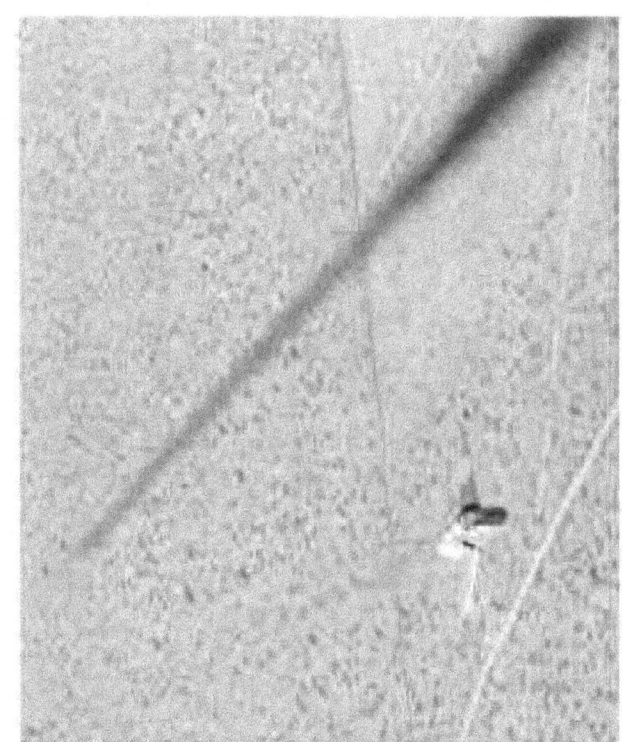

Used to retain instrumented knuckle assembly

>> Generate slack using power

>> Select idle and speedbrakes

>> Standby for a big jolt!

Lessons Learned

- Simulation configuration control is important
- Design for "real life" human factors
- Rope aerodynamic forces need to be modeled better
- Some rope elasticity is beneficial
- Low tow works well for less efficient "gliders"
- Don't underestimate the uses of GPS
- Simulation paid huge dividends
- Multi-organizational teams need clearly defined roles, responsibilities, and joint goals.
- Don't ignore handheld data

167

Index

About the Author

Tom Tucker is a writer who has a special interest in topics relating to invention. He is the author of *Touchdown: The Development of Propulsion Controlled Aircraft at NASA Dryden* (Washington, DC: Monographs in Aerospace History # 16, 1999) and *Brainstorm: The Stories of Twenty American Kid Inventors* (New York: Farrar, Straus & Giroux, 1995, revised 1998). His next nonfiction work, "Bolt of Fate: Benjamin Franklin and his Electric Kite Experiment" is due out from Public Affairs Books in the winter of 2001. He has published in many periodicals and also has written about baseball, including a baseball short story featured in *Sports Illustrated.* He is an instructor at Isothermal Community College in Spindale, NC. He attended Harvard College and Washington University in St. Louis, earning his B.A. and later an M.A. on a Woodrow Wilson Fellowship at Washington University.

Monographs in Aerospace History

Launius, Roger D., and Gillette, Aaron K. Compilers. *The Space Shuttle: An Annotated Bibliography.* (Monographs in Aerospace History, No. 1, 1992).

Launius, Roger D., and Hunley, J.D. Compilers. *An Annotated Bibliography of the Apollo Program.* (Monographs in Aerospace History, No. 2, 1994).

Launius, Roger D. *Apollo: A Retrospective Analysis.* (Monographs in Aerospace History, No. 3, 1994).

Hansen, James R. *Enchanted Rendezvous: John C. Houbolt and the Genesis of the Lunar-Orbit Rendezvous Concept.* (Monographs in Aerospace History, No. 4, 1995).

Gorn, Michael H. *Hugh L. Dryden's Career in Aviation and Space.* (Monographs in Aerospace History, No. 5, 1996).

Powers, Sheryll Goecke. *Women in Flight Research at the Dryden Flight Research Center, 1946-1995* (Monographs in Aerospace History, No. 6, 1997).

Portree, David S.F. and Trevino, Robert C. Compilers. *Walking to Olympus: A Chronology of Extravehicular Activity (EVA).* (Monographs in Aerospace History, No. 7, 1997).

Logsdon, John M. Moderator. *The Legislative Origins of the National Aeronautics and Space Act of 1958: Proceedings of an Oral History Workshop* (Monographs in Aerospace History, No. 8, 1998).

Rumerman, Judy A. Compiler. *U.S. Human Spaceflight: A Record of Achievement, 1961-1998* (Monographs in Aerospace History, No. 9, 1998).

Portree, David S.F. *NASA's Origins and the Dawn of the Space Age* (Monographs in Aerospace History, No. 10, 1998).

Logsdon, John M. *Together in Orbit: The Origins of International Cooperation in the Space Station Program* (Monographs in Aerospace History, No. 11, 1998).

Phillips, W. Hewitt. *Journey in Aeronautical Research: A Career at NASA Langley Research Center* (Monographs in Aerospace History, No. 12, 1998).

Braslow, Albert L. *A History of Suction-Type Laminar-Flow Control with Emphasis on Flight Research* (Monographs in Aerospace History, No. 13, 1999).

Logsdon, John M. Moderator. *Managing the Moon Program: Lessons Learned from Project Apollo* (Monographs in Aerospace History, No. 14, 1999).

Perminov, V.G. *The Difficult Road to Mars: A Brief History of Mars Exploration in the Soviet Union* (Monographs in Aerospace History, No. 15, 1999).

Tucker, Tom. *Touchdown: The Development of Propulsion Controlled Aircraft at NASA Dryden* (Monographs in Aerospace History, No. 16, 1999).

Maisel, Martin D.; Demo J. Giulianetti; and Daniel C. Dugan. *The History of the XV-15 Tilt Rotor Research Aircraft: From Concept to Flight.* (Monographs in Aerospace History #17, NASA SP-2000-4517, 2000).

Jenkins, Dennis R. *Hypersonics Before the Shuttle: A History of the X-15 Research Airplane.* (Monographs in Aerospace History #18, NASA SP-2000-4518, 2000).

Chambers, Joseph R. *Partners in Freedom: Contributions of the Langley Research Center to U.S. Military Aircraft in the 1990s.* (Monographs in Aerospace History #19, NASA SP-2000-4519).

Waltman, Gene L. *Black Magic and Gremlins: Analog Flight Simulations at NASA's Flight Research Center.* (Monographs in Aerospace History #20, NASA SP-2000-4520).

Portree, David S.F. *Humans to Mars: Fifty Years of Mission Planning, 1950-2000.* (Monographs in Aerospace History #21, NASA SP-2002-4521).

Thompson, Milton O. *Flight Research: Problems Encountered and What They Should Teach Us.* (Monographs in Aerospace History #22, NASA SP-2000-4522).

Those monographs still in print are available free of charge from the NASA History Division, Code ZH, NASA Headquarters, Washington, DC 20546. Please enclosed a self-addressed 9x12" envelope stamped for 15 ounces for these items.